实用
土木与建筑工程
专业英语

Practical English for
Architecture & Civil Engineering

主 编◎汪德华
副主编◎（按姓氏笔画排序）
闫一江 / 赵 斌 / 崔 莉

浙江大学出版社

图书在版编目（CIP）数据

实用土木与建筑工程专业英语 / 汪德华主编. —杭州：
浙江大学出版社，2013.5（2022.6重印）
ISBN 978-7-308-10964-2

Ⅰ.①实… Ⅱ.①汪… Ⅲ.①土木工程－英语②建筑
工程－英语 Ⅳ.①H31

中国版本图书馆 CIP 数据核字（2013）第 000504 号

实用土木与建筑工程专业英语

主　编　汪德华
副主编　闫一江　赵　斌　崔　莉

责任编辑	李　晨
封面设计	续设计
出版发行	浙江大学出版社
	（杭州市天目山路 148 号　邮政编码 310007）
	（网址：http://www.zjupress.com）
排　　版	杭州青翊图文设计有限公司
印　　刷	嘉兴华源印刷厂
开　　本	787mm×1092mm　1/16
印　　张	18.25
字　　数	608 千
版 印 次	2013 年 5 月第 1 版　2022 年 6 月第 4 次印刷
书　　号	ISBN 978-7-308-10964-2
定　　价	46.00 元

版权所有　翻印必究　　印装差错　负责调换

浙江大学出版社市场运营中心联系方式：0571－88925591；http://zjdxcbs.tmall.com

作者简介

汪德华 浙江大学宁波理工学院教授,"东西方语言文化与传播中心"副主任。宁波市外文翻译学会理事。研究方向:英汉语言与文化,英语教学法。在国内外学术刊物上发表论文40余篇,主编教材《建筑工程专业英语》《中国和英美国家习俗文化比较》等7本,主持完成了浙江省社科联,河南省社科联重点科研课题及民生课题10余项,其中"我国加入WTO对外语教育的影响与对策研究"获河南省社科联重点调研课题一等奖;河南省人民政府实用社会科学优秀成果三等奖。"高校实施创新教育,培养创新人才问题的研究"获河南省教育科学研究优秀成果一等奖。

闫一江 宁波市城建设计研究院结构工程师。湖南大学土木工程学院毕业,硕士。宁波市土木建筑工程学会会员。研究方向:配筋砌体剪力墙结构。在专业杂志上发表《混凝土小型空心砌块墙体数值模拟及刚度分析》等多篇论文,参与了宁波市多项重大工程项目的设计工作。多项工程设计被评为浙江省、宁波市优秀设计二、三等奖。

赵 斌 浙江大学宁波理工学院外语分院教师,从事高校外语教学十余年,曾多年从事对外贸易并驻海外承担国际承包工程口、笔译工作,为"双师型"教师,负责浙江省教育厅课题一项、宁波市级课题两项并获2008年宁波市教育科研优秀成果二等奖。

崔 莉 浙江大学宁波理工学院讲师,硕士,主要从事大学英语教学和中西方文化研究,主持浙江省社会科学界联合会研究课题1项、宁波市哲学社会科学规划课题1项,并在国内学术期刊上发表多篇论文。

前 言

为适应我国高等教育发展的新形势,深化教学改革,提高教学质量,满足新时期国家和社会对人才培养的需要,2004年初,教育部颁布了大学英语课程教学要求(试行),拉开了全国大学英语课程改革的序幕。大学英语教学作为一门基础课程,其重要意义众所周知,但两年英语学习之后,学生的语言能力在实际应用中表现得并不理想,主要原因之一是没有解决好四年不断线的课程设置问题。因此,各高等学校根据实际情况,按照课程要求和本校的大学英语教学目标设计出各自的大学英语课程体系,将综合英语类、语言技能类、语言应用类、语言文化类和专业英语类等必修课程和选修课程有机结合,以确保不同层次的学生在英语应用能力方面得到充分的训练和提高。不同专业的学生在修完基础阶段的英语课程后,可根据自身兴趣和需要自由选择提高阶段课程的学习,以满足个人和社会发展的需求。

本书为适应高等院校建筑工程类专业英语的教学需要而编写,本书注重学习者的学习兴趣、专业知识和认知水平,倡导自主、体验和实践的学习方式,强调"在学中练,在练中学,在学中用,在用中学"的学习理念。既注重知识结构的完整性,又突出所选材料的趣味性。全书课文以建筑工程为主线,系统地介绍该专业所包括的基本内容。全书共14个单元,内容涉及土木建筑工程各个方面,系统地介绍该专业所包括的基本内容:①专业介绍;②建筑材料;③建筑设计;④建筑结构;⑤建筑基础;⑥建筑墙体;⑦屋顶建筑;⑧建筑工具及方法;⑨建造程序;⑩建筑物的功能;⑪建筑测量;⑫建筑保温与防潮;⑬建筑设计比例;⑭建造后期问题。每个单元包括五个部分:第一部分为热身准备,第二部分为指导实践,第三部分为能力扩展,第四部分为建筑商务,第五部分为专业阅读。五个环节,环环相扣,集说、写、读、译为一体。一到四部分着力培养学生对专业知识的表述能力,其中练习形式多样,轻松活泼,设计的句子简单,实用,易记,且附有插图。第五部分的阅读,设计(A),(B)两篇阅读文章,选材新,文体符合一般英语原版书刊的格式,难易适中,涉及面广。每篇文章后附有生词、注释、练习。书后附有每单元练习的参考答案。

《实用土木与建筑工程专业英语》具有以下特点:

1. 直观生动:教材内容编排进行全新的尝试,体例新颖。每单元附有相关内容的插图,使教学内容更加直观,有利于提高学生的学习兴趣及课程的教学质量。

2. 难度适中:内容和语言设计充分考虑普通高校土建专业学生的实际水平,根据语言技能和专业知识的不同方面编排教学内容,内容实用,易学易记。

3. 突出应用:教材体系严密完整。弱化理论,强化实践内容,侧重技能传授。各章节内容围绕学生在该领域将要面对并应该了解和掌握的问题展开,使学生既掌握英语语言技能又掌握该专业技能。

4. 内容丰富:本教材涉及土建专业活动的各个环节,同时还包括语言、文化、经济和商务等方面的知识。使学生提高专业英语阅读能力的同时还扩大知识面。

5. 使用广泛:该书可作为本科大学英语四级后的专业英语教材,亦可供高职高专土建专业的学生使用,还可以供土木与建筑工程在职人员业务培训用。

本书由汪德华规划全书结构、选编材料及担任主审。赵斌、崔莉参与了材料的选编和改写,并编制了课后练习。闫一江承担了该书的选材、校对、审核,对书中的专业插图进行了绘制,并对本书的文字排版做了大量的工作。由于编者水平有限,书中难免出现一些错误,谨请读者谅解,并恳请广大读者和同行专家提出宝贵批评意见。

编 者

2012 年 6 月 16 日

CONTENTS

UNIT 1   INTRODUCTION ··· 1
UNIT 2   BUILDING MATERIALS ·· 15
UNIT 3   HOUSE DESIGNING ·· 31
UNIT 4   STRUCTURE ··· 46
UNIT 5   FOUNDATION ··· 62
UNIT 6   WALLS ·· 79
UNIT 7   ROOFS ·· 96
UNIT 8   TOOLS AND METHODS ··· 112
UNIT 9   PROCESS ·· 128
UNIT 10 FUNCTIONS OF BUILDING ······································ 146
UNIT 11 MEASUREMENTS ··· 170
UNIT 12 THERMAL AND MOISTURE PROTECTION ················· 184
UNIT 13 PROPORTIONS IN DESIGN ······································ 200
UNIT 14 POST-CONSTRUCTION PROBLEMS ··························· 217
Keys to the Exercises from Unit 1 to Unit 14 ··························· 232
APPENDIX ··· 264
REFERENCES ··· 283

UNIT 1   INTRODUCTION

Part 1   Warm-up Activities

1. Here are some examples of basic forms, read out and learn them by heart:

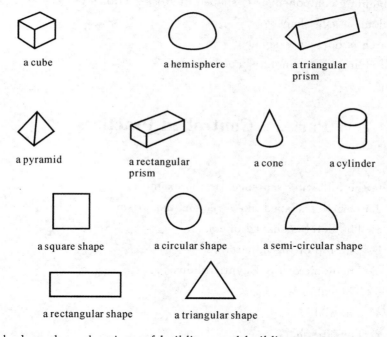

2. Now look at these drawings of buildings and building components:

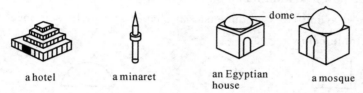

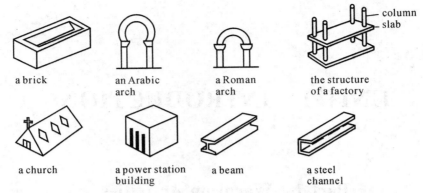

3. Now answer these questions about the drawings in exercise 2:
a) Which building is pencil-shaped?
b) Which building component is T-shaped in cross-section?
c) Which building component is C-shaped in cross-section?
d) Which dome is egg-shaped?
e) Which arch is horseshoe-shaped?
f) Which building has diamond-shaped windows?

Part 2 Controlled Practices

1. Complete the following sentences as the example:
Example: The brick is shaped like a rectangular prism.
       a) The hotel is shaped like a...
       b) The top of the minaret...
       c) The dome of the Egyptian house...
       d) The column...
       e) The slab...
       f) The church...
       g) The power station building...

2. Look at these drawings of two-dimensional shapes and make sentences using the phrases given in the table below:

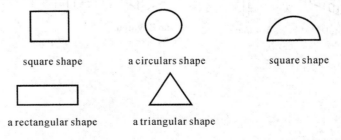

Example:

The cross-section of a square prism is square in shape.

The longitudinal section of a square prism is rectangular in shape.

The cross-section of the	brick hotel top of the minaret	is	square circular semi-circular	in shape.
The longitudinal section of the	column church		rectangular triangular	

3. Study, read and follow:

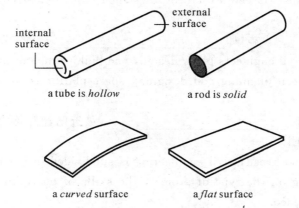

a tube is *hollow*   a rod is *solid*

a *curved* surface   a *flat* surface

The power station building is hollow. It has five flat external surfaces.

Now describe these buildings and components in a similar way:

a) The church

b) The slab

c) The column

d) The mosque

e) The steel beam

Part 3 Further Development

Civil Engineering and Civil Engineers

The following statements are about civil engineering and civil engineers. Read the English sentences and fill in the blanks with proper translation.

1) Civil engineering, the oldest of the engineering specialties, is the planning, design, construction, and management of the built environment. This environment includes all structures built according to scientific principles, from irrigation and drainage

systems to rocket-launching facilities.

土木工程,最古老的工程专业,是建筑环境的_____。这个建筑环境包括从_____到_____所有根据科学原理建造的结构物。

2) Civil engineers build roads, bridges, tunnels, dams, harbors, power plants, water and sewage systems, hospitals, schools, mass transit, and other public facilities essential to modern society and large population concentrations.

土木工程师修建_____以及现代化社会和大量人口集中的地方所必需的其他公共设施。

3) They also build privately owned facilities such as airports, railroads, pipelines, skyscrapers and other large structures designed for industrial, commercial, or residential use.

他们也修建_____设施,如_____设计的其他大型建筑。

4) In addition, civil engineers plan, design, and build complete cities and towns, and more recently have been planning and designing space platforms to house self-contained communities.

此外,土木工程师_____,最近已经开始规划和设计_____以容纳独立的(科研)团体。

5) Because it is so broad, civil engineering is subdivided into a number of technical specialties. Depending on the type of project, the skills of many kinds of civil engineer specialists may be needed.

因为土木工程的范围太广,所以它被细分为许多技术专业。根据_____,就需要土木工程师专家的_____。

6) When a project begins, the site is surveyed and mapped by civil engineers who locate utility placement-water, sewer and power lines. Geotechnical specialists perform soil experiments to determine if the earth can bear the weight of the project. Meanwhile, structural specialists use preliminary data to make detailed designs, plans and specifications for the project.

当一项工程开始时,_____要勘测现场并绘图,他们还要确定水管、污水管道和电线的实用布置。_____要做土工试验以确定该土壤是否能承受这项工程的重量。同时,用初始资料来做工程的详细设计、规划和说明书。

7) Environmental specialists study the project's impact on the local area: the potential for air and groundwater pollution, the project's impact on local animal and plant life, and how the project can be designed to meet government requirements aimed at protecting the environment.

_____要研究工程对当地区域的影响:_____,工程对当地动植物的影响,以及工程怎样设计才能_____保护环境的_____。

8) Transportation specialists determine what kind of facilities are needed to ease the burden on local roads and other transportation networks that will result from the completed project.

4

_____要确定需用_____来减轻由完工的工程产生的荷载_____带来的压力。

9) Supervising and coordinating the work of these civil engineer specialists, from the beginning to the end of the project, are the construction management specialists Based on information supplied by the other specialists, construction management civil engineers estimate quantities and costs of materials and labor, schedule all work, order materials and equipment for the job, hire contractors and subcontractors, and perform other supervisory work to ensure the project is completed on time and as specified.

从工程开始到结束,_____监督并协调土木工程专家们工作。根据_____提供的信息,施工管理土木工程师要估计材料、劳动力的数量和成本,安排所有的工作,_____,_____,雇用承包商和转包人,以及_____以确保工程能按照说明按时完工。

10) Throughout any given project, civil engineers make extensive use of computers. Computers are used to design the project's various elements (computer-aided design, or CAD) and to manage it.

对于_____,土木工程师都能_____。计算机_____并进行管理。

Part 4　Business Activities

International Travel

1. 文化与指南(Culture and Directions)

国外承包工程,涉及许多国际旅行(international traveling)的问题,因此,需要学习一些旅途用语,了解一些乘飞机的常识。

最好在起飞前1个小时到机场。然后到相关航空公司的机场服务台(check-in counter)凭机票(ticket)和护照(passport)领取登机卡(boarding card),并办理托运行李手续。过安检时要将登机牌、身份证交给安检员,安检员审核没问题后会在登机牌上面盖章。然后过安检门,随身带的物品从安检门旁的X光安检机传送过去,人要从安检门通过。安检没问题就进候机厅候机。听到登机广播后,到登机口将登机牌交给服务人员检验,持登机牌上飞机。登机牌上标明有你的位置,数字代表第几排,每排的座位是按A、B、C、D、E、F排列的,找到你的位置坐下,扣上安全带,起飞前关掉手机。在飞行时,航空公司有免费饮料派发,长航线会有免费餐食供应,短航线派发点心。航行中,空服员会发一份入境登记表或海关申报表,应该早一点填妥。下飞机后,要先到机场移民局,交上旅客入境卡,移民局官员在检查你的护照和签证并询问一些问题之后,若没发现问题,就会让你入境。接下来你需要到领取行李处(baggage claim area)领取行李。最后,再到海关(customs)上交海关申报单,办理入关手续。

2. 情景会话（Situational Conversations）

(1) Booking An Airplane Ticket

China Electrical Construction Company (CECC) is awarded a contract for a power project abroad. The project manager's assistant is phoning to book an airplane ticket.
(A: the project manager's assistant; B: the ticket agent)

B: East China Airlines. How can I help you?

A: I'd like to make a reservation for Alger, please.

B: When do you plan to leave?

A: I'll set off on the 23rd of next month.

B: We have seats available on a flight leaving Shanghai at 9 o'clock in the morning and arriving in Alger at 11 o'clock in the evening of the same day. Is that all right?

A: Sounds great.

B: Would you prefer first class, business or economy?

A: Economy, please.

B: May I have your name?

A: Zhang Shan and another person is Lin Qiang.

B: OK. I've booked you on East China Airlines Flight Number EC135 leaving at 9 o'clock, June 23rd and arriving in Alger at 11 o'clock.

A: All right. Thank you very much.

B: How would you like to pay for the ticket?

A: By credit card. Can I give you the details now and then you can send me the tickets through the post?

B: Certainly.

A: Thank you.

B: You are welcome.

(2) Check-in at the Airport

(A: the project manager's assistant; B: the airport clerk)

B: (At the check-in counter) Good morning. Ticket and passport, please.

A: Good morning. We're in a group. Here are the tickets and passports.

B: How many people in the group?

A: Two.

B: Any seat preference?

A: We'd prefer the window seats in the non-smoking section.

B: How many pieces of baggage do you want to check?

A: Two for each of us. By the way, what's the baggage allowance?

B: Seventy pounds for each passenger and the weight limit for each piece is forty-five pounds. Put these four big cartons and that suitcase on the scale then, one by one.
A: I hope they're not overweight.

(After weight all the things)
B: No, you're OK. Now, let me put on the baggage tags.
B: Your boarding cards and passports.
A: Thanks. What's the boarding gate number?
B: Gate seven. Have a good flight, gentlemen.
A: Thanks a lot.

(3) On Board the Plane

(A: *the project manager*; B: *the project manager's assistant*; C: *Steward*)

C: (At lunch time) What would you like to have, sir? We have fried chicken, fish and beef.
B: Fish, please.
C: How about a drink, coke, orange juice or beer?
B: Do you have tea?
C: Sorry, We don't serve tea at lunch.
B: Then, a coke, please.
C: (To Mr. Lin, who is sitting next to Mr. Zhang) And you, sir?
A: Nothing right now. I'm feeling a bit airsick.
C: I'll see if I can get you some tablets for airsickness. Just a moment, please.
A: Thank you very much.

(4) Going through Customs

(A: *the project manager*; B: *airport inspector*)

B: Good afternoon. May I see your passport, please?
A: Of course. Here you are.
B: Thank you. What is the purpose of your visit—business or pleasure?
A: Business. We're going to Alger to execute a construction project.
B: Fine. Is this all your luggage?
A: Yes, that's all my luggage, one suitcase and one bag.
B: Do you have anything to declare?
A: I guess not.
B: Well, would you mind opening your suitcase?

A: Oh, not at all.
B: Thanks.

(*Zhang opens his suitcase. The inspector is inspecting the suitcase, and now he looks at a bag.*)

B: What's inside the bag?
A: That's my laptop computer. Do I have to pay any duty on it?
B: No, it's duty-free.
A: By the way, I'm carrying four packs of cigarettes for my own use. Are they dutiable?
B: No, goods for personal use rather than commercial use are not subject to duty. And they are within the limit.
A: Good. And thanks for the information.
B: All right. Here's your passport.
A: Is that all the customs formalities?
B: Yes. You're though now. Have a pleasant stay.
A: Thank a lot.

New Words and Expressions：

available *a.*	可以得到的,可以利用的
destination *n.*	目的地
stewardess *n.*	航空小姐
airsick *n.*	晕机
passport *n.*	护照
tablet *n.*	药片
construction project	建筑项目,工程项目
customs *n.*	海关
declare *v.*	申报
belongings *n.*	所有物,物品
check-in	登记,报到
flight number *n.*	航班号
airport terminal *n.*	机场候机大厅
A. C. (airline clerk)	航空公司职员
baggage *n.*	行李
allowance *n.*	许可量,限额
suitcase *n.*	手提箱,衣箱
overweight *v.*	超重的
boarding card	登机卡
boarding gate	登机门

Notes to the Text:

1. to execute a project：承建一项目，如：We're going to execute a power project.
2. declaration slip：海关申报单。也可以说 declaration form, customs statement。
3. personal belongings：个人随身物品。也可以说 personal effects。
4. check-in counter：办理登机手续的服务台，又称作 check-in desk。
5. baggage allowance：免费托运的行李限额。如果超过限额，则超出的部分要付运费（pay for excess baggage）。
6. baggage tag：行李标签，也可以说 baggage claim tag，上面填有托运人的姓名以及托运到的地点。
7. boarding card：登机卡（牌），也可以说 boarding pass。

Post-Reading Exercises:

1. Do you know any regulations for taking a plane? Tick the statements you think are true.

   (　　)1) Passengers can only hand-carry a small bag aboard the plane.
   (　　)2) Children under 5 needn't buy a ticket.
   (　　)3) People who haven't got an ID card cannot board the plane.
   (　　)4) Passengers mustn't carry dangerous articles such as compressed gases, weapons, explosives, corrosives, or inflammables onto the plane.
   (　　)5) Passengers can smoke during the flight.
   (　　)6) Passengers mustn't use mobile telephones on flights because they will interfere with the plane's electronic equipment.
   (　　)7) Passengers must check in 30 minutes before departure on international flights.
   (　　)8) Passengers must fasten their seat belts during take-off or landing.
   (　　)9) The airline needn't accept responsibility for delays due to bad weather.
   (　　)10) Passengers can transfer their tickets to others without going to the booking office.

2. Translate the following sentences into Chinese:

   1) For your safety, all electronic devices, such as mobile phones, computers, CD players, MP3/MP4, game players must be turned off during take-off and landing.
   2) Please make sure your baggage is in the over-head bin or under the seat in front of you.
   3) Please keep the aisle and the exits free of baggage.
   4) Attention, please. We'll be landing at Los Angeles Airport in 30 minutes' time. For your own safety, please make sure that your seat belt is well fastened.
   5) Please take your passport, flight ticket, relevant immigration forms and all your belongings with you and go through the immigration formalities in the terminal.

Part 5　Extensive Reading

(A) Houses

A house is a building that provides home for one or more families. Its main function is to provide shelter, but a house usually serves many more purposes. It is a center of family activities, a place for entertaining friends, and a source of pride in its comfort and appearance.

Many private houses built in the United States are designed by builders, rather than by professional architects. For this reason, family houses are most easily classified by the type of layout or the floor plan, not by reference to architectural styles. Here are some examples of commonly used styles of family homes in the United States.

Cape Cod. The Cape Cod house is usually small and has a sharply sloping roof. Many Cape Cods have three or four rooms on the ground floor and two small rooms on the second floor. In the modified Cape Cod, the second floor is made larger by decreasing the slope of the roof in the rear and by adding dormers, or gabled extensions with windows, in the front. The Cape Cod originated in colonial New England.

Two-Story House. Before World War I, the usual one-family house had two full stories, as well as a basement and an attic. This type of house is probably still the most common house, and many new ones are being built following this design, It usually has a square floor plan and a central hall, and it is often called a Colonial house.

Ranch House. The ranch house is long and low and all the rooms are at ground level. Some ranch houses have no basement. The ranch house originated in the West and Southwest of the United States and became popular in the building boom that followed World War II.

Split-level House. The split-level house has two or more levels, separated from each other by half flights of steps. The kitchen and family room may be on one level. The living room on a second level, and the bedrooms on a third level. This house was developed in the building boom that followed World War I, and it is probably the most popular type being built.

Attached Houses. Houses for two families are often called duplex houses, because they have two stories. One family may live on each story, or each family may have both upstairs rooms and downstairs rooms, with the two parts of the house separated by a wall through the middle.

Row houses. In towns and cities, there are often rows or two-story houses attached to each other. These are called row houses. Each unit in a row may be a one-family or two-family house.

The basic function of housing is to provide shelter from the element. In the mid-1960s, a most important value in housing was sufficient space both inside and outside. A majority of families preferred single-family homes on about half an acre of land, which would provide space for spare-time activities outdoors. Many families preferred to live as far away from the center of a metropolitan area as possible, even if the wage earners had to travel some distance to their work. About four out of ten families preferred country housing to suburban housing because their chief aim was to get far away from noise, crowding, and confusion of the city center. The accessibility of public transportation had ceased to be a decisive factor in housing because most workers drove their cars to work; people were chiefly interested in the arrangement and size of rooms and the number of bedrooms.

However, people today require much more than this of their housing. A family moving into a new neighborhood will be interested to find out about the district they are moving to. In regard to safety, health and comfort, people's requirements have increased. A family may also ask how near the housing is to churches, schools, stores, the library, a movie theatre and the community center. It has become more and more popular to choose the new location for a family home according to the living standard of the neighborhood.

New Words and Expressions:

entertain	v.	招待,款待,请客
layout	n.	布局,安排,设计
slope	n.	斜面,斜坡;倾斜,斜度
rear	n./a.	后部,尾部,背面;后方
dormer	n.	天窗
colonial	a.	殖民的
attic	n.	阁楼
ranch house	n.	平房建筑
boom	n.	托架;起重臂;吊杆;弦杆
duplex house	n.	双联住宅
acre	n.	英亩
decisive	a.	决定性的

Post-Reading Exercises:

1. Translate the following phrases into Chinese:
1) a one-family house
2) to provide shelter
3) classified by the type of layout or the floor plan
4) a sharply sloping roof

5) separated by half flights of steps

6) the number of bedrooms

2. Decide whether the following statements are true(T) or false(F) according to the passage given above.

1) A house is a building that provides home for one or more families.　　　(　　)

2) In the United States all private houses are designed by professional architects.
　　　　　　　　　　　　　　　　　　　　　　　　　　　　　　　　　(　　)

3) Before World War I, the usual one-family house had 3 full stories, as well as a basement and an attic.　　　(　　)

4) A split-level house has two or more levels, separated from each other by a full flight of steps.　　　(　　)

5) Houses attached to each other in towns and cities are called row houses.　(　　)

6) The basic function of housing is to provide shelter e.g. against wind, rain and sun.　　　(　　)

7) In the mid-1960's, many families preferred to stay in the center of a metropolitan area, even if it was noisy and dirty.　　　(　　)

8) Nowadays, sufficient space both inside and outside is the most important value for housing.　　　(　　)

3. Answer the following questions:

1) What is the main function of a building? Give some examples of additional functions.

2) Who can design buildings in the United States?

3) What styles of family homes do you know according to the reading passage?

4) What is the name of a small house with three or four rooms on the ground floor, two small rooms on the second floor and a sharply sloping roof?

5) What is the name of the house with all the rooms on the ground level?

6) What is the name of a house with several levels that are separated by half flights of steps, kitchen and family room on the ground floor, living room on a second and bedrooms on a third level.

7) Which house has added dormers or gabled extensions with windows in the front?

8) In what way have the requirements for family housing changed during the last decades?

(B) A Career in Architecture

Why would a career in architecture attract you? Well, if you are the kind of person who is curious about your surroundings, then you might just be interested in learning how to improve them. As an architect you would have the power and the responsibility to shape the environments in which people spend their daily lives. This makes architecture one of the most influential professions in today's society. We are living in a rapidly changing world, and so we need people with the imagination to create the buildings and cities our society needs to keep pace with progress. If you are some one who is excited by change, then you could grasp the opportunity to build the future the way you want it. Now read about the career profile of Isabel Allen, a female architect from Great Britain, who is working as editor of the architectural magazine *Architects' Journal*.

"After completing an economics degree at Manchester University, I took a degree in architecture at the University of Westminster. This was followed by working as an architectural assistant in Singapore, where I worked on various projects including the Open University building in Singapore, a nursery school in China, and various housing developments in China and Singapore. I returned to England to take a post graduate diploma in architecture at South Bank University, I completed the course in 1996 and immediately joined the *Architects' Journal*, as Buildings Editor. My job was to decide which buildings to publish, source photographs and articles or write them myself.

In 1999 I became editor of the *Architects' Journal*. I am responsible for a team of 12 editorial staff members. My job changes according to the day of the week. On Monday morning I have a meeting with people who represent the commercial side of the magazine to discuss how we are doing financially. There is a commercial aspect to my job-talking to the advertising team about advertising opportunities, meeting advertisers and making sure that we don't spend more than we can afford.

Monday and Tuesday are our press days and the whole team tends to concentrate on that week's issue. On the other days I have meetings with architects or advertisers and I plan and prepare features for future issues-we are working on twelve different issues at any given time. Every Friday the editorial team meets to discuss future issues and any problems, which have come up during the week.

There are various public duties, which go with the job, such as acting as a judge for architectural competitions, visiting schools of architecture, or giving interviews to newspapers, television or radio when there is an architectural story in the news. But my main responsibility is to make sure that the content of the *Architects' Journal* is always interesting, intelligent and accurate."

New Words and Expressions:

architect	*n.*	建筑师
influential	*a.*	有影响的
profile	*n.*	侧面,剖面;剖面图,外形
diploma	*n.*	证书
assistant	*n.*	助手,助理,副手
source	*n.*	源点
commercial	*a.*	商业的
accurate	*a.*	精确的

Post-Reading Exercises:

1. Translate the following phrases into Chinese:

1) a career in architecture

2) to keep pace with progress

3) to complete a degree at the university

4) to take a post graduate diploma in architecture at the university

5) to concentrate on the week's issue to take part in architectural competitions

2. Decide whether the following statements are true(T) or false(F) according to the passage given above.

1) If you are the kind of person who is always interested in your surroundings, then a career in architecture will attract you. (　　)

2) As an architect, you are able to change the shape of buildings and consequently influence the daily lives of people. (　　)

3) Isabel Allen studied architecture at many universities around the world such as the South Bank University in Great Britain and the Open University in Singapore.

4) In 1996, she became editor of the *Architects' Journal* and since then she is responsible for a team of 12 editorial staff members. (　　)

5) The task of a Buildings Editor is to decide which buildings to publish, to check the source of photos and articles or sometimes even write some articles. (　　)

6) Monday and Tuesday are the so called press days, which mean that the whole team is forced to work faster and to concentrate on that week's issue. (　　)

7) Her job as an editor of the *Architects' journal* includes taking part in architecture competitions with her own design projects. (　　)

8) Today, her main responsibility is to take care of the content of the *Architects' Journal* and to provide the readers with an interesting, intelligent and accurate magazine.
(　　)

3. Answer the following questions:

1) What are main requirements for a career in architecture? Add some examples according to your own experience.

2) What is the role of an architect nowadays in our rapidly changing world?

3) What was her major at Manchester University?

4) At which universities did she study?

5) Has she ever worked as an architect designing buildings?

6) What different kind of job experiences does she have?

UNIT 2   BUILDING MATERIALS

Part 1   Warm-up Activities

1. Learn the following by heart and make out their Chinese names:

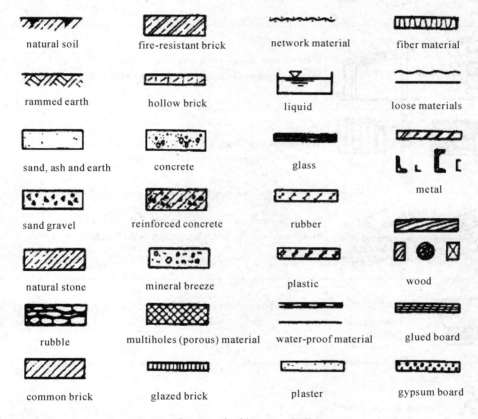

Common building material

2. Study these sentences and read out:

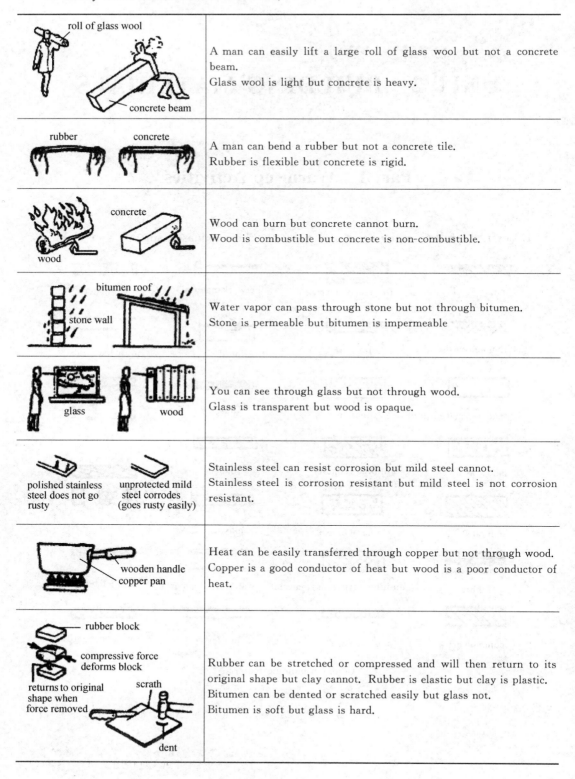

roll of glass wool / concrete beam	A man can easily lift a large roll of glass wool but not a concrete beam. Glass wool is light but concrete is heavy.
rubber / concrete	A man can bend a rubber but not a concrete tile. Rubber is flexible but concrete is rigid.
wood / concrete	Wood can burn but concrete cannot burn. Wood is combustible but concrete is non-combustible.
stone wall / bitumen roof	Water vapor can pass through stone but not through bitumen. Stone is permeable but bitumen is impermeable
glass / wood	You can see through glass but not through wood. Glass is transparent but wood is opaque.
polished stainless steel does not go rusty / unprotected mild steel corrodes (goes rusty easily)	Stainless steel can resist corrosion but mild steel cannot. Stainless steel is corrosion resistant but mild steel is not corrosion resistant.
wooden handle copper pan	Heat can be easily transferred through copper but not through wood. Copper is a good conductor of heat but wood is a poor conductor of heat.
rubber block / compressive force deforms block / returns to original shape when force removed / scrath / dent	Rubber can be stretched or compressed and will then return to its original shape but clay cannot. Rubber is elastic but clay is plastic. Bitumen can be dented or scratched easily but glass not. Bitumen is soft but glass is hard.

3. Make sentences about four other properties of materials from this table:

Steel Stone Glass Wool Brick	has the property of	good sound insulation. good thermal insulation. high compressive strength. high tensile strength.
This means		it can resist high compressive forces. it can resist high tensile forces. it does not transmit heat easily. it does not transmit sound easily.

4. Answer the following questions:

Why is glass used for window panes?

Because glass is _____ .

Why is glass wool used to keep the heat in hot-water tanks?

Because glass wool has the property of...

Why is some steel covered with a thin layer of zinc?

Because zinc is...

Why are some fire doors covered with asbestos sheets?

Because asbestos is...

Why are some metal sheets formed into a corrugated shape?

Because the corrugated shape makes the sheet _____ .

Why is concrete used for the columns of a building structure?

Because...

Part 2   Controlled Practices

1. Study these diagrams. Match the letters A—H in the diagrams with the sentences below:

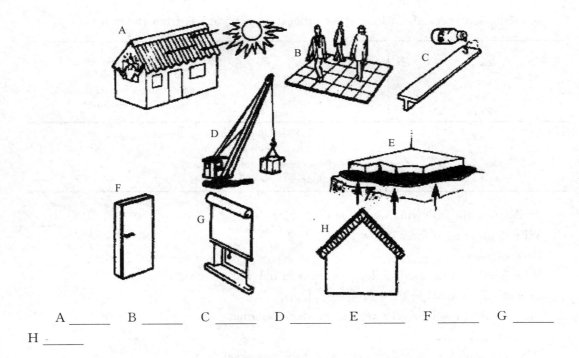

A _____ B _____ C _____ D _____ E _____ F _____ G _____ H _____

2. Now complete these sentences with the matching material properties:

1) The polythene membrane can prevent moisture from rising into the concrete floor. This means that polythene is _____.

2) The T-shaped aluminum section can resist chemical action, i.e. aluminum is _____.

3) The stone block cannot be lifted without using a crane. This means that stone is _____.

4) The corrugated iron roof cannot prevent the sun from heating up the house, i.e. iron is _____.

5) Glass wool can help to keep a house warm in the winter and cool in the summer; i.e. glass wool is _____.

6) The ceramic tiles on the floor cannot be scratched easily by people walking on them. This means that ceramic tiles are _____.

7) Asbestos sheeting can be used to fireproof doors. In other words asbestos is _____.

8) Black cloth blinds can be used to keep the light out of a room, i.e. cloth is _____.

UNIT 2 BUILDING MATERIALS

3. Complete this table with notes about all the materials described above:

Material	Availability	Use	Properties	Problems/Durability
cane leaves vine bamboo palm-fronds	warm-humid zones			
grass				
hardwoods and softwoods				
Earth				
Concrete				

4. Copy and complete this table by putting ticks in the boxes to show the functions of the components:

	Function of components		
Form of material	Structural support only	Space dividing only	Both structural support and space dividing
blocks			
sheets			
rods			

5. Now decide whether these statements are true(T) or false(F). Correct the false statements.

   a) Rod materials can be used for both dividing space and supporting the building.

   b) Concrete can be used as a block material, a sheet material and a rod material.

   c) Steel is used for frame construction because it has high tensile strength and low compressive strength.

   d) The sheet materials, which act as space dividers in a frame construction building, can be very light because they do not support structural loads.

   e) Mass construction buildings are light whereas planar construction buildings are heavy.

6. Now complete these sentences:

   a) A copper tube is an example of a _____, because it is...

   b) A concrete block is an example of a _____, because...

   c) A steel stanchion is an example of a _____, because...

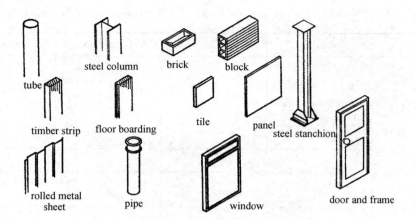

Part 3　Further Development

1. Make sentences according to the examples：

1) What is the feature about?

e.g. What is the feature about this material?（这些材料的特性是什么？）

Your sentences：_____

2) answer our purpose

e.g. The construction material answers our purpose satisfactorily.（这种建筑材料能满足我们的需要。）

Your sentences：_____

3) The average...of...is

e.g. The average traffic fuel（gasoline）consumption of this lorry is 0.3 liter per kilometer(l/km).（这台货车的平均行车柴油（汽油）耗量为每公里0.3公升。）

Your sentences：_____

4) come from

e.g. This special oil comes from the "SHELL" company（CALTEX，MOCBIL，GULF，ESSO，CASTROL，BP）.这种特种油来自"壳牌"公司（加德士、飞马、海湾、埃索、卡斯特罗、英国石油公司。）

Your sentences：_____

5) compare with

e.g. Cast iron cannot compare with steel in tensile strength.（铸铁在抗拉强度上比不上钢。）

Your sentences：_____

2. Read the sentences both in English and in Chinese. Try to memorize the words for construction materials.

1) Cement steel and timber are the most important construction materials used in civil engineering. 水泥、钢材和木材是土建工程中最重要的建筑材料。

2) Typical structural steel shapes include beams, channels, angles and tees. 典型的结构型钢包括工字钢、槽钢、角钢和丁字钢。

3) There are four broad classifications of steel: carbon steels, alloy steels, high-strength low-alloy steels and stainless steels. 钢材大致可分为四类，即：碳素钢、合金钢、高强度低合金钢和不锈钢。

4) Copper, zinc, lead, aluminum, bronze and brass are nonferrous metals or alloys. 铜、锌、铅、铝、青铜和黄铜都是有色金属或合金。

5) This alloy is mainly composed of element chromium and nickel (titanium, vanadium, manganese). 这种合金主要由元素铬和镍(钛、钡、锰)组成。

6) The standards GB and YB provide the method of testing for materials in our country just like the standard ASTM in America. 在我国，GB(国标)和YB(冶标)规定了材料的试验方法，正如美国的 ASTM 标准一样。

7) We have asbestos(rubber, plastic, glass, paint) products of all kinds. 我们有各种石棉(橡胶、塑料、玻璃、油漆)制品。

8) Bolt(screw, nut, stud, spring washer, pin, ball bearing, roller bearing) is the most commonly used machine part. 螺栓(螺钉、螺帽、双头螺栓、弹簧垫圈、销、滚珠轴承、滚柱轴承)是最常用的机械零件。

Part 4　Business Activities

Settling Down

1. 文化与指南(Culture and Directions)

初到工程施工所在国，首先解决的就是居住(housing)问题。少数先遣人员可以先住宾馆。很多时候出国前都要预订好宾馆房间，预订宾馆的方式也是多种多样的。电话、上网、信函、电传都是可以的，但最常用的还是电话预订。

但由于后续大批施工人员的到来，住宾馆既不经济也不方便，公司需要建造施工营地(construction camp)。通常可选的方案是委托当地一家住宅建筑公司承建(subletting the construction camp)。建好之前，一般会租一所离工地较近的房子。租房前应先了解一下当地租房的行情。包括每月租金(rental)、租期(lease)、是否要押金(security deposit)、房子里有哪些设施以及设施的使用费是否包含在租金中，等等。如果对该房条件满意，需与房东(landlord/landlady)签订一个租赁协议(sign a lease agreement)。

2. 情景会话 (Situational Conversations)

(1) Making A Room Reservation

(*Before they go abroad, the assistant of the project manager phones a local hotel to book a room for the group, who are going to Alger to execute a construction project.*)
(*A: Assistant of the project manager; B: Receptionist*)

B: Advance Reservations. How can I help you?

A: Hello, I'd like to reserve a room for two, I wonder if you still have any rooms available.

B: When do you plan to stay here?

A: From the 23rd to the end of this month.

B: Wait a minute. Let me see. Yes, we still have some rooms available.

A: That would be fine. What's the rate per night?

B: $150 a night.

A: What services come with that?

B: That includes breakfast and there is a colour television, an IDD telephone and a computer with Internet access.

A: OK, thank you. I'll take it.

B: Very good. What's your name, please?

A: Li Mei—L-I, M-E-I.

B: Thank you. Can you give me a contact number, please?

A: 89212239.

B: Thank you very much. We look forward to seeing you.

(*Checking in at the hotel*)

A: I have booked a double room in your hotel.

B: Let me check. Oh, yes, Room 907 is reserved for you. Would you please sign your name in the registration form?

A: I wonder if it's possible to reserve a small conference room in your hotel. Tomorrow we'll have a meeting there.

B: Yes, we have a conference room which holds about 10 people.

A: By the way, how long does the laundry service take?

B: Laundry is collected at nine in the morning and will be returned in the afternoon of the same day.

A: I see.

B: Here's your key. Your room is on the 9th floor, The porter will take your luggage up. Have a good day.

A: Thanks a lot.

(2) Subletting the Construction Camp

(A: *The project manager*; B: *Manager of a local company which specializes in residential buildings*)

B: Hello, Mr. Yang. Very pleased to meet you in person.

A: Very pleased to meet you, too, Mr. Howell.

B: Let's get down to business, Mr. Yang. You asked me on the phone whether we would like to bid for a construction camp to accommodate eighty men. Could you put it in more detail, please?

A: As you know, the camp will mainly be built for the accommodation of about eighty Chinese men who will be working for the project. It includes double occupancy staff units, four man occupancy units for workers and cooks, a kitchen and dining unit, shower bath units, office units, a conference room and a recreation unit.

B: So the camp will accommodate about eighty men. Do you have any specific requirements?

A: I would like to hear your recommendation.

B: In my opinion, there's a kind of prefabricated house which is most suitable for such a camp. It's easy and economical to transport, fast to erect and very convenient to dismantle for either relocation or disposal when the whole project is completed.

A: Sounds fine. What's it made of?

B: Light concrete slabs.

A: Good. Would you please give us a quotation for such a camp on a turn-key basis as soon as possible?

B: All right. By the way, who will be responsible for leveling the ground for the camp site?

A: We will do that. You will be responsible for the water and power supply. One more thing, this is a duty-free project. So everything imported for it is duty-free. Please take it into consideration in your quotation.

B: In that case, our quotation will be much lower.

A: We really appreciate that.

B: When do you want the camp to be completed?

A: Within sixty days, starting from our notice to commence the work.

B: OK, Mr. Yang. I hope we can be given the opportunity to work for you and I believe we will surely do a good job.

A: I hope so, too.

New Words and Expressions：

construction crew	施工人员（总称）
specification n.	技术规范，（产品）规格
accommodation n.	居住，膳宿条件
sublet v.	分包
residential building	住宅楼
accommodate n.	容纳，提供（住宿）
shower bath	淋浴
recreation unit	娱乐室
prefabricated a.	预制的
erect v.	安装，装配
dismantle v.	拆除
relocation n.	搬迁
disposal n.	处理
light concrete slab	轻混凝土板
quotation n.	报价
finalize v.	确定
level the ground	平整场地
duty-free a.	免税的
commence the work	开工

Notes to the Text：

1. bid for：为……而投标。例如：bid for an irrigation project（为一个灌溉工程项目而投标）。与该词组有相同意思的还有 tender for。

2. prefabricated house：装配式房屋，活动房屋。prefabricated 这个词在工程实施中常常使用，与它搭配的词很多，例如：prefabricated construction（装配式施工），prefabricated pile（预制桩），prefabricated unit（预制构件）。

3. on a "turn-key" basis：以"交钥匙"的方式。

Post-Reading Exercises：

"Who will be responsible for leveling the ground for the camp site?"（哪一方负责平整营地的场地？）

"You'll be responsible for the water and power supply."（但你方应负责供水供电。）

Often the word "responsible" is used in specifying and clarifying the responsibilities between the parties to a contract during a negotiation. Now you try doing the following translation：

1) A：We need all the data on hydrological and subsurface conditions before we start the job.

B：We could provide relevant data to you，(但你方应对……负责) _____ your own interpretation of them.

2) A: We require that you rectify the damages which happened to the works on the 15th of this month at your own cost.

B: We don't think that(我方不应为此损坏负责)＿＿＿＿＿＿. As stipulated in the contract,(工程的某一部分一旦移交给业主)＿＿＿＿＿＿, the contractor shall not have the obligation to protect it.

3) A: If we accept the subcontractor nominated by you, ＿＿＿＿＿＿(我们是否就其或其雇员的行为向你方负责?)

B: Yes, as the general contractor, ＿＿＿＿＿＿＿＿＿＿(你们应为引起的一切后果负责).

Part 5 Extensive Reading

(A) Building Materials

Materials for building must have certain physical properties to be structurally useful. Primarily, they must be able to carry a load or weight, without changing shape permanently. When a load is applied to a structure member, it will deform. For example, a wire will stretch or a beam will bend.

However, when the load is removed, the wire and the beam come back to the original positions. This material property is called elasticity. If a material were not elastic and a deformation were present in the structure after removal of the load, repeated loading and unloading eventually would increase the deformation to the point where the structure would become useless. All materials used in architectural structures, such as wood, steel, aluminum, reinforced concrete and plastics, behave elastically within a certain defined range of loading. If the loading is increased above the range, two types of behavior can occur: brittle and plastic. In the former, the material will break suddenly. In the latter, the material begins to at a certain load(yield strength), ultimately leading to fracture. As examples, steel exhibits plastic behavior and stone is brittle. The ultimate strength of a material is measured by the stress, at which failure(fracture) occurs.

A second important property of a building material is its stiffness. This property is defined by the elastic modulus, which is the ratio of the stress(force per unit area), to the strain(deformation per unit length). The elastic modulus, therefore, is a measure of the resistance of a material to deformation under load. For two materials of equal area under the same load, the one with the higher elastic modulus has the smaller deformation. Structural steel, which has an elastic modulus of 30 million pounds per square inch(psi) or 2,100,000 kilograms per square centimeter, is 3 times as stiff as aluminum, 10 times as stiff as concrete, and 15 times as stiff as wood or burnt clay or slate, and concrete blocks.

Masonry. Masonry is essentially a compressive material; it cannot withstand a tensile force, that is, a pull. The ultimate compressive strength of bonded masonry depends on the strength of the masonry unit and the mortar. The ultimate strength will vary from 1,000 to 4,000 psi (70 to 284 kg/sq cm), depending on the particular combination of masonry unit and mortar used.

Timber. Timber is one of the earliest construction materials and one of the few natural materials with good tensile properties. Hundreds of different species of wood are found throughout the world, and each species exhibits different physical characteristics. Only a few species are used structurally as framing members in building construction. In the United States, for instance, out of more than 600 species of wood only 20 species are used structurally. These are generally softwoods, because of their abundance and the ease with which their wood can be shaped.

Steel. Steel is an outstanding structural material. It has a high strength on a pound-for-pound basis when compared to other materials, even though its volume-for-volume weight is more than ten times that of wood. It has a high elastic modulus, which results in small deformations under load. It can be formed by rolling into various structural shapes such as I-beams, plates, and sheets; it also can be cast into complex shapes; and it is also produced in the form of wire strands and ropes for use as cables in suspension bridges and suspended roofs.

Aluminum. Aluminum can be formed into a variety of shapes; it can be extruded to form I-beams, drawn to form wire and rods, and rolled to form foil and plates. Aluminum members can be put together in the same way as steel by riveting, bolting and (to a lesser extent) welding. Apart from its use for framing members in buildings and prefabricated housing, aluminum also finds extensive use for window frames and for the skin of the building in curtain wall construction.

Concrete. Concrete is a mixture of water, sand, gravel and Portland cement. Crushed stone, manufactured lightweight stone, and seashells are often used in lieu of natural gravel. Portland cement, which is a mixture of materials containing calcium and clay, is heated in a kiln and then pulverized.

New Words and Expressions:

permanently	*ad.*	持久地,不变地
deform	*v.*	使变形
deformation	*n.*	畸形
architectural	*a.*	建筑上的
aluminum	*n.*	(化)铝
fracture	*n.*	折断
brittle	*a.*	易碎的,脆弱的

ratio	n.	比例，比率
clay	n.	黏土，泥土
slate	n.	石板
masonry	n.	砖石建筑
tensile	a.	张力的，可伸长的
mortar	n.	泥浆，胶泥
timber	n.	木材，横木
conifer	n.	针叶树
cable	n.	缆索，电缆
suspension	n.	悬吊，悬浮，悬挂
extrude	v.	挤压，撑出
prefabricate	v.	预制
gravel	n.	碎石，沙砾层
pulverize	v.	粉碎，磨成粉
rivet	n.	铆钉
lieu	n.	场所
kiln	n.	干燥炉，炉

Post-Reading Exercises:

1. Answer the following questions:

1) What building materials are mentioned in the passage? Make a list of them.

2) If the loading is increased above the range, what can occur?

3) Do all species of wood can be used as building material? If not, what kind of wood suit?

4) Which of the materials can be extruded to form I-beams, drawn to form wire and rods, and rolled to form foil and plates?

2. Fill in the blanks:

1) Materials for building must have certain physical properties, such as _____.

2) All materials used in architectural structures behave _____ within a certain defined range of loading.

3) _____ is one of the earliest construction materials and one of the few natural materials with good tensile properties.

4) _____ is essentially a compressive material, which cannot withstand a tensile force.

5) _____ has a high elastic modulus, which results in small deformations under load.

6) _____ members can be put together in the same way as steel by riveting.

7) _____, which is a mixture of materials containing calcium and clay, is heated in a kiln and then pulverized.

8) When the load is removed, the wire comes back to the original positions. This material property is called _____.

(B) Wood Panel Products

The development of processed products was a major step for the timber industry. Today the precious raw material wood can be used in a more efficient way than just for solid wooden boards and pillars.

Some examples of such processed wood materials are laminated wood and wood panel products.

Laminated wood is made of many small strips of wood that are glued together or joined together with mechanical fastenings to form a large piece of wood. It is still solid wood, but it does not depend on the natural height of a tree any longer. Laminated wood is available as big structural elements in any desired size including curves and angles.

Wood panel products are big timber elements made of several thin layers of veneers that are glued together. According to the order in which the layers of veneers are glued together we can differ between different panel products such as plywood panels, composite panels and non-veneered panels. A popular example of non-veneered panels is the so-called OSB board, which means Oriented Strand Board. All wood panel products require less labor for installation than solid boards because they are bigger and hence fewer pieces must be handled.

It is a main advantage that wood panel products are equal in strength in their 2 main directions. In comparison, solid wood is several times stronger parallel to the grain than perpendicular to grain because of the wood cell structure.

There are 3 ways of slicing the veneer: Rotary slicing, plain slicing and quarter slicing.

For rotary slicing, the timber logs are soaked in hot water to soften the wood, and then each is rotated in a large lathe against a stationary knife that peels off a continuous strip of veneer, much as paper is unwound from a roll(Look at the picture given above). The strip of veneer is clipped into sheets that pass through a drying kiln where in a few minutes their moisture content is reduced to roughly 5 percent. The sheets are then assembled into larger sheets, repaired as necessary with patches glued into the sheet to fill open defects, graded and sorted according to quality. A machine spreads glue onto the veneers, as they are laid atop one another in the required sequence and grain orientations. The glued veneers are transformed into plywood in presses that apply elevated temperatures and pressures to create dense, flat panels. The panels are trimmed to size, sanded as required, and grade stamped before shipping.

The most economical method is rotary slicing, which is used for veneers for structural wood panels. The finest figures are produced by quarter slicing, which results in a very close grain pattern with prominent rays. The grain figure produced by rotary slicing is extremely broad and uneven. For better control of the grain figure in face veneers, the veneers are plain sliced or quarter sliced.

New Words and Expressions:

pillar	n.	支柱、栋梁
laminate	v.	碾压
curve	n.	曲线
veneer	n.	胶合板饰面
installation	n.	装置设备
parallel	n.	平行线
slice	v.	成薄片
atop	prep.	在……的顶上
clip	n.	夹子
dense	a.	密集的
elevate	v.	举起
rotary	a.	转动的
prominent	a.	突出的
defect	n.	缺陷

Post-Reading Exercises:

1. Decide whether the following statements are true(T) or false(F) according to the passage given above.

1) Today there are many ways to use wood as building material. (　　)

2) The size of laminated wood depends on the natural size of a tree in the forest. The taller the tree, the bigger the wooden board. (　　)

3) Wood panel products are still solid wood products. They have several thin layers of veneers that are glued together, but the core in the middle consists of a solid wooden board. (　　)

4) Oriented strand boards are non-veneered panels and are made of several layers of veneers that are glued together. (　　)

5) After drying in a kiln, the rotary sliced sheets of veneer still have roughly 5 percent moisture content. (　　)

6) Plywood is made of glued, rotary sliced veneers and has the shape of dense, flat panels. (　　)

7) The most economical method of slicing is robot slicing, which is used fox veneers for structural wood panels. (　　)

8) To get a smooth surface texture with a grain figure, face veneers are plain sliced or quarter sliced. (　　)

2. Answer the following questions:

1) In what ways is wood used for construction purposes? Give 3 examples.
2) How is laminated wood made? Describe the process.
3) What do laminated wood elements look like?
4) How do wood panel products look like?
5) What are the 3 different slicing methods? Describe them briefly.
6) Why are timber logs for rotary slicing soaked in hot water?
7) How are open defects repaired?
8) Compare the different slicing methods, and find advantages and disadvantages.

UNIT 3　HOUSE DESIGNING

Part 1　Warm-up Activities

1. Look at this drawing of the exterior of a single-story house and read out:

The sidewalls are perpendicular to the front wall.

The front wall is parallel to the back wall.

The ground slab foundation extends beyond the perimeter walls.

The roof of the house is removed by cutting along line AB that is parallel to the floor.

Now look at a cut-away view of the interior of the same house.

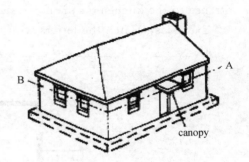

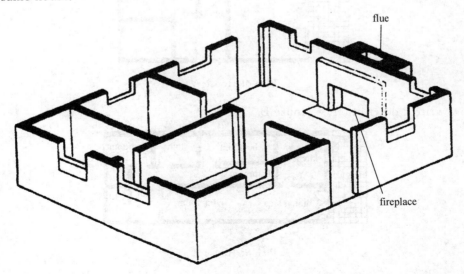

A view looking straight down on a cut-away view of the interior a building is known as a plan.

Looking north, the living room is on the right of the house.

Looking north, the bedroom is to the left of the living room.

The kitchen is next to the living room.

Viewed from the front, the kitchen is behind the bedroom(or the bedroom is in front of the kitchen).

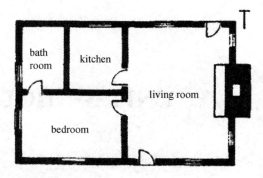

2. Read this description of House A:

House A is a single-storey building with a square-shaped plan. It contains seven rooms. The entrance which is located on the south side leads into a hall. On the left of the hall is the living room and beyond that in the north-west corner is the dining area. The kitchen is adjacent to the dining area. A terrace is situated outside the living room on the west side. A toilet is located in the centre of the house. Access to the toilet is from the hall. The two bedrooms are located on the east side with a bathroom between them. There is also an entrance to the kitchen on the north side.

Exercises: Match the letters with the names of the areas:

a) _____ b) _____ c) _____ d) _____ e) _____ f) _____ g) _____
h) _____ i) _____

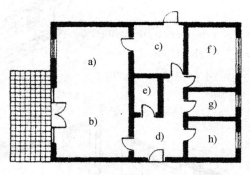

Plan of House A

Now write a description of House B.

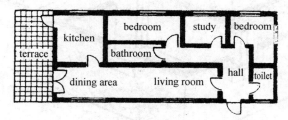

Plan of House B

Part 2 Controlled Practices

1. Study these plans of a two-story house:

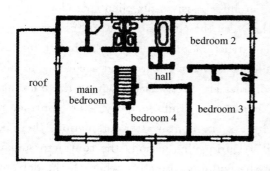 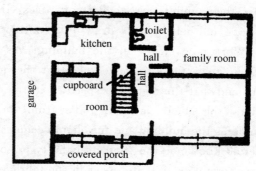

First floor plan of House B   Ground floor plan of House B

2. Now decide whether these statements are true(T) or false(F). Correct the false ones.

a) The dining room is located under the main bedroom.
b) A hall is located in the center of the first floor.
c) There are three adjacent bathrooms on the first floor.
d) There is a toilet between the kitchen and the dining room.
e) Bedroom 2 is situated over the family room.
f) There is a cupboard under the stairs.
g) Bedrooms occupy most of the ground floor.
h) Viewed from the front, the dining room is on the left of the entrance.
i) Viewed from the rear, the living room is behind the family room.
j) From the garage, you pass through the living room to enter the family room.
k) The entrance is situated at the bottom of the stairs.
l) The kitchen and family room are located on either side of the toilet.
m) A door in the garage leads to the kitchen.

3. Read and find a word or an expression in the following passage which means:

1) to be given the job of designing a building
2) to offer to a client for his consideration
3) to combine into a whole lot
4) to offer to do some work at a certain price
5) to look at the building work in detail at regular intervals
6) an interval of time after the building has been finished during which the contractor is responsible for correcting any faults in it
7) to have complete ownership of the building

When an architect receives a commission for a building, he meets the client and discusses his requirements. After visiting the site, the architect draws up preliminary plans, and together with a rough estimate of the cost, submits them to the client for his approval. If the client suggests changes, the architect incorporates them into the final design which shows the exact dimension of every part of the building. At this stage, several building contractors are invited to bid for the job of constructing the building. When they submit their tenders or prices, the architect assists his client in selecting the best one and helps him to draw up a contract between the client and the contractor.

Work now starts on the building. As construction proceeds, the architect makes periodic inspections to make sure that the building is being constructed according to his plans and that the materials specified in the contract are being used. During the building period, the client pays the bills from the contractor. Subsequently, the contractor completes the building and the client occupies it. For six months after completion, there is a period known as the "defects liability period". During this period, the contractor must correct any defects that appear in the fabric of the building. Finally, when all the defects have been corrected, the client takes full possession of the building.

4. Complete this flow diagram:

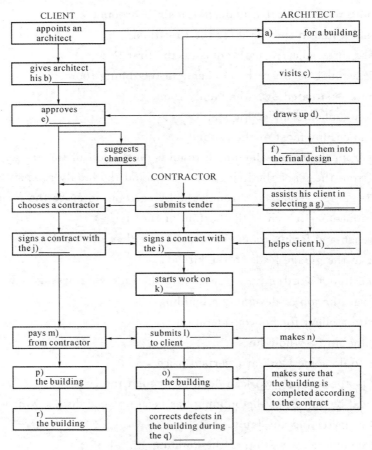

Part 3 Further Development

1. Read the expressions for drawings both in English and in Chinese. Try to memorize them.

the contents of drawings	图纸目录
structural detail drawing	结构大样图
structural plane diagram	结构平面图
building detail drawing	建筑大样图
general building plan	建筑总平面图
general construction master plan	施工总平面图
construction drawing	施工图
perspective drawing	透视图
foundation plan	基础平面图
piled foundation drawing	桩基础图
concrete foundation drawing	砼基础图
rubble foundation drawing	毛石基础图
reinforced concrete foundation drawing	钢筋砼基础图
prefabricated components drawing	预制（装配式）构件图
drawings of door and windows	门窗图

2. Read the sentences both in English and in Chinese. Try to memorize them.

1) This is a plot plan(general layout, general arrangement, detail, section, erection, flow sheet, PID, assembly, civil, electrical, control and instrumentation, projection, piping, isometric) drawing. 这是一张平面布置（总平面、总布置、细部、剖面、安装、流程、带仪表控制点的管道、装配、土建、电气、自控和仪表、投影、配管、等距）图。

2) That is a general(front, rear, side, left, right, top, vertical, bottom, elevation, auxiliary, cut-away, birds eye) view. 那是全视（前视、后视、侧视、左视、右视、顶视、俯视、底视、立视、辅助、内部剖视、鸟瞰）图。

3) The information to be placed in each title block of a drawing include: drawing number, drawing size, scale, weight, sheet number and number of sheets, drawing title and signatures of persons preparing, checking and approving the drawing. 每张图纸的图标栏内容包括：图号、图纸尺寸、比例、重量、页码和张数、图标以及图纸的制图、校对、批准人的签字。

4) There are various types of lines on the drawing such as: border lines, visible lines, invisible lines, section lines, central lines, size lines, break lines, phantom lines. 图上有各种形式的线条，诸如：边框线、实线、虚线、剖面线、中心线、尺寸线、断裂线、假想线。

5) Please give us a copy of this information(technical specification, instruction,

manual, document, diagram, catalog). 请给我们一份这个资料(技术规程、说明书、手册、文件、图表、目录样本)的复印本。

6) We have not received this drawing(instruction book, operation manual), please help us to get it. 我们还未收到这张图纸(说明书、操作手册),请帮助我们取得。

7) Please explain the meaning of this abbreviation(mark, symbol) on the drawing. 请解释图上这个缩写(标记、符号)的意义。

3. Translate the following sentences into Chinese:

1) Please send us further information about this item.

2) We want additional information on this.

3) A working drawing must be clear and complete.

4) How many drawings are there in the set?

5) Is this a copy for reproduction?

6) What is the edition of this drawing?

7) Is this drawing in effect?

8) Is this a revised edition?

Part 4 Business Activities

Possession of the Site

1. 文化与指南(Culture and Directions)

工程的施工管理是承包商履行合同过程中非常重要的一部分,也是企业能否盈利的重要因素之一。科学合理的施工管理是企业在市场竞争中能否胜出的先决条件。施工管理涉及面很广,包括施工项目的组织机构、人员设备管理、工期管理、质量管理及施工安全、劳动保护和环境保护等方面。为了使施工管理达到科学合理的目标,实施合同的项目部应注意以下几点:(1)应掌握现代化的管理理论和方法;(2)尽可能地熟悉合同文件、图纸、规范以及当地政府主管部门和办事流程;(3)尽可能地了解当地法律法规;(4)技术和管理团队应当技术过硬而且具有团队精神;(5)应妥善归档保存合同的相关文件、资料、信函等。

2. 情景会话(Situational Conversations)

(1) Discussing Organization Chart

The management of constructing a project is involved in many activities such as the overall planning, coordination and control of it from inception to completion. So, many discussion and meetings are needed between the client and the contractor in order to produce a functionally and financially viable project. Now the client's consultant engineer and the contractor's project engineer are discussing the Construction Method Statement.

(A: *The client's consultant engineer*; B: *The contractor's project engineer*)

A: Today we would like to discuss with you from Chapter 3 of the Construction Method Statement. Let's see the Organization Chart first.

B: This Organization Chart is a typical organization for large industrial plants we have performed.

A: How many people compose the project management?

B: Totally around 200. The top management has three people only. They are Project Manager, Deputy Project Manager and Chief Engineer. Under project management there are five departments: Works, Technical, Procurement, Administration and Quality Departments.

A: What about the functions of each department?

B: The Works Department is the largest one and responsible for all construction sections and teams. The Deputy Project Manager is also the Manager of Works Department. The Technical Department is in charge of all matters about drawings and designing of temporary facilities.

A: Which department deals with the subcontract matters?

B: It also belongs to the responsibility of the Technical Department. The Procurement Department is in charge of material and construction equipment supply and warehouse. The Administration Department is in charge of all matters related to finance and public relationship. The Quality Department is a special agency for it is controlled by not only Project Management but also the Headquarter of the company.

A: I think the Organization Chart should be supported by an introduction to the key persons of the site management.

B: The key persons for the project management are qualified engineers with at least ten years experience in construction field especially for power plant works. We will select our best staff to be the responsible persons for this important project.

A: Yes, it should be. I would like to remind you that as long as these key staffs are approved to work for the project, they should not be removed from the site without our permission.

B: No, they won't be. We understand the importance of keeping capable persons for such an important project.

(2) Discussing Construction Machine Schedule

(A: *The client's consultant engineer*; B: *The contractor's project engineer*)

A: Now let's discuss the Chapter 5: Construction Machine Schedule. We quite appreciate that you intend to mobilize all necessary construction equipment for the works. But there is still some shortage of equipment to be applied on the site; for example, the concrete batching plants.

B: We proposed to use 4 Nos. of batching plant with a capacity of $25m^3$/hour(25 cubic meters per hour). That means we have $100m^3$ concrete per hour capacity. The maximum concrete order is $2000m^3$ per day.

A: But you can not guarantee all plants will operate very well during the peak period without a spare set. And also your plants do not look up to the standard. They take too large space and the measuring for materiel is not accurate enough. Since this is key equipment, we suggest you use some new and modern typed batching plants.

B: We will consider this point and find a solution very soon. Probably we will employ 3 Nos. of $40m^3$/h batching plant instead.

A: That will be better. Who is the manufacturer of the batching plant? Can you let us have the specification of the batching plants?

B: The same manufacturer with the original style: the Construction Machine Manufacturer belongs to our corporation which is a qualified and famous producer in China. We will provide a detailed specification of the plants.

A: Another thing we would like to point out about the construction equipment is that you need at least one set of power generator in case the power supply from the city net switched off since power supply in the city is not stable.

B: OK. We will prepare two sets of power generators, one in construction area and the other in the living area.

A: One more thing we would suggest is that the capacity of the cranes seems not adequate. For such heavy weight of a single steel truss and beam, one set of 150t (tone) crawler crane is absolutely necessary.

B: We plan to use two sets of 50t crawler working together instead of one set of 150t.

A: We are not sure it is a good proposal. Please study this carefully and provide more details of how you are lifting heavy steel elements in your Construction Method Description.

B: OK. We will consider it and study if there is any problem of doing so by our method. We will prepare an alternative method in case there is any risk existed.

New Words and Expressions:

consent *n./v.*   同意
statement *n.*   陈述、说明、方案
chart *n.*   (施工)图表、排行榜
compose *v.*   组成,构成
be in charge of   负责,主管

subcontract *n./v.* 分包,转包
procurement *n.* 采购
supervise *v.* 监督
interference *n.* 介入,干预
mobilize: *v.* 调配,使用
capacity *n.* 能力,容量、容积
stable *a.* 稳定的
crawler crane 履带式塔吊

Notes to the Text:

1. organization chart:（公司）组织机构图。
2. project management:项目管理部门。
3. The Quality Department is a special agency:质量(监控)部是一个特殊的部门。
4. key persons:主要人士。
5. select our best staff:选配我方最好的工作人员。
6. construction Machine Schedule:施工设备(明细)表。
7. concrete batching plants:混凝土搅拌机。
8. maximum concrete order:最大混凝土(浇筑)量。order 此处意为"数量"。
9. peak period:（用电、交通等）高峰时段。
10. in case the power supply from the city net switched off since power supply in the city is not stable:以防备市区供电网断电因为城里供电不稳定。since 意为"因为"。

Post-Reading Exercises:

Translate the following sentences into Chinese:

1) Under project management there are five departments: Works, Technical, Procurement, Administration and Quality Departments.

2) Now let's discuss the Chapter 5: Construction Machine Schedule. We quite appreciate that you intend to mobilize all necessary construction equipment for the works. But there is still some shortage of equipment to be applied on the site; for example, the concrete batching plants.

3) But you can not guarantee all plants will operate very well during the peak period without a spare set. And also your plants do not look up to the standard. They take too large space and the measuring for materiel is not accurate enough. Since this is key equipment, we suggest you use some new and modern typed batching plants.

Part 5 Extensive Reading

(A) Architectural Drawings

From his appearance on earth to the end of his cave-dwelling days, man had very little need for architectural drawing. When man for the most part abandoned cave dwelling, cave-like shelters appeared. They were simple in design and their construction matched the technology of that day. Drawings were not necessary.

The appearance of formal shapes with remarkably accurate dimensions, however, suggests there was a need for architectural drawing. It is difficult to imagine achieving such precision work with the limited construction technology available without some form of pictorial planning and documentation. With more complex buildings being built, it became necessary to develop more elaborate drawing methods. In the last few hundred years, architectural drawing has evolved into several general types. These range from concept sketches to intricate details drawn to scale. Design sketches are rough drawings that are used as "idea sketches", made to explore concepts that will be refined at a later date. They may appear crude to the casual observer, but a closer study of sketches drawn by talented designers will usually reveal a theme and sensitivity containing the elements of good design. It is the purpose of these drawings to establish such elements. Design sketches have changed very little from the earliest known examples to those of today's architects.

Usually, clients are not trained in grasping concepts from rough sketches. Therefore, a more picture-like drawing, known as a presentation drawing, is required to explain a proposed building to the prospective owner. Such a drawing may also be required for the banker or loan company who will be asked to lend money to build the project and for other persons or committees that must first understand, then approve or disapprove of the proposal.

Presentation drawings are usually necessary for commercial work and custom homes, while homes built on a speculative basis with no owner committed usually do not require this preparation. Often, modern design presentation drawings are characterized by realistic features, such as shades, shadows, people, trees, plantings and automobiles. The techniques for making these drawings have changed over the years as drawing technology has advanced and attitudes have changed concerning what parts of the building or setting are most important.

Once the design has been accepted, drawings must be prepared to guide the builders of the project. These are called "working drawings". They are precisely drawn and include plan views, elevations and details, all with dimensions and notes.

Another type of architectural drawing is one that is completed after the construction of a building. It is called "an as-built drawing" if it contains dimensions and other technical data. It may be for use in further technical work, for maintenance of the building, or it may be used for publication. Architectural drawing is the means of graphic communication used within the professions and businesses that are concerned with the design and construction of buildings. The building profession falls generally into two categories, residential and nonresidential. Nonresidential includes commercial, institutional industrial, recreational, and other types of buildings that are not houses. This group is sometimes referred to simply as commercial. Small commercial buildings built with residential methods are called "light construction".

Residential work includes multi-family apartments, condominiums town houses and single family buildings. Although state laws vary, typically it is a legal requirement that new nonresidential and multifamily buildings be designed by registered architects. Employees of a registered architect need not be licensed, however. Usually it is not required to have an architectural registration seal (license) to design and detail single family and duplex residences. To become a licensed architect one typically earns a degree in architecture from an accredited university, serves an apprenticeship under a registered architect, and then passes a lengthy examination. The architect is a major source but not the only generator of architectural drawings. Architects work closely with and often coordinate the efforts of many consultants, who perform design and drawing work. Engineers, interior designers, landscape architects, acoustical consultants and hospital specialists are some of the professionals who are associated with architects and who also employ architectural drafters.

Notes to the Text:

1. It is difficult to imagine achieving such precision work with the limited construction technology available without some form of pictorial planning and documentation:在所能获得的施工技术十分有限的情况下,要完成精确度如此高的工程而没有用图片形式表现的设计是很难想象的(precision 表示"精确性", available 表示"所获得的",修饰 technology)。

2. With more complex buildings being built:随着更复杂建筑物的修建。

3. concept sketches:初步概念图。

4. drawn to scale:按比例绘制。

5. that are used as"idea sketches", made to explore concepts that will be refined at a later date:用作"初步概念图",供探讨初步设想用,以便以后进一步细化。

6. usually reveal a theme and sensitivity containing the elements of good design:通常揭示设计的主题思想和孕育优秀设计成分的敏感性。

7. a presentation drawing:表现图。

8. while home built on a speculative basis with no owner committed usually do not require this preparation:根据市场预测而不是根据买主委托而建造的住宅则通常不要求绘制这种表现图(speculative:推测的;committed:受托付的)。

9. Attitudes have changed concerning...: attitude concerning... have changed: 因后置定语 concerning 比较长,按英语习惯应放在谓语 have changed 的后面。

10. working drawings: 施工图。

11. as-built drawings: 竣工图。

12. institutional: (学校、医院等)事业单位的。

13. condominiums: 共有公寓房(房产为私有,庭园等为公有)。

New Words and Expressions:

cave-dwelling	n.	穴居
dimensions	n.	大小,程度,体积,范围
precision	n.	精确性
pictorial	a.	用图片、照片表示的
elaborate	a.	详尽而复杂的
intricate	a.	错综复杂的
sketch	n.	草图
speculative	a.	思索的,推测的
presentation	n.	介绍,描述,显示
elevations	n.	正视图
graphic	a.	生动的
duplex	a.	复式的,双联的
drafter	n.	起草人

Post-Reading Exercises:

1. Choose the one that best completes the sentence:

1) Today's skyscrapers are big buildings, whose construction _____ the technology of the day.

  A. equals   B. pairs   C. matches   D. reaches

2) The extinction of many species of animals on earth _____ the need for environmental protection.

  A. supports   B. offers   C. demands   D. suggests

3) Modern technology has developed _____ machines, which have eased man's life.

  A. careful   B. elaborate   C. simple   D. needed

4) An as-built drawing is one, which is not to be _____ at a later date.

  A. refined   B. drawn   C. copied   D. published

5) As we know, most of the maps are drawn to _____.

  A. scale   B. model   C. proportion   D. size

6) Sometimes, one has to draw a conclusion on a(n) _____ basis.

  A. perspective   B. speculative   C. instinctive   D. persuasive

7) Architectural drawing is a form of _____ communication.
   A. graphic   B. body   C. language   D. verbal
8) Nonresidential buildings are sometimes _____ as commercial.
   A. regarded   B. referred to   C. looked upon   D. mistaken
9) A(n) _____ architect is one who has passed the necessary examination required by local authority.
   A. registered   B. competent   C. professional   D. institutional
10) Students graduating from an _____ university are highly respected by the society.
   A. acclaimed   B. aspired   C. accredited   D. advanced

2. Answer the following questions:
1) Why were architectural drawings not necessary when cave-like shelters appeared?
2) What are the basic types of architectural drawings?
3) What is the purpose of design sketches?
4) What realistic features do presentation drawings contain?
5) What is the function of an as-built drawing?
6) In what way does nonresidential work differ from residential?
7) What is required of new multifamily and nonresidential buildings in the U. S. ?
8) What are the requirements for becoming a registered architect?

(B) Complexity and Contradiction

Robert Venturi is famous for his book *Complexity and Contradiction* (1966) in which he called attention to the importance of Baroque architecture in contradiction to rectilinear Modernism. In his later book *Learning from Las Vegas* (1972), Venturi and Denise Scott Brown proposed the idea of the "decorated shed" (ordinary American architecture) and proposed it could be perceived as artistic inspiration. Venturi argued for an architecture that was neither pure nor picturesque, but which made the most of complexities, contradictions, ambiguities and paradoxes-qualities which he thought were more understanding of the times. Venturi said that architecture did not have to be "heroic and original". Instead it was "OK" to look back upon rich architectural history for inspiration and references.

Venturi calls the house "a small house on a large scale", thus Venturi's mother sits in a chair with a pot of flowers at her feet to help create a sense of scale. This house for his mother allowed Venturi to build his ideas about complexity and contradiction. Historical references were used boldly with a bit of humor. Venturi violates the rules of Modernism with his gabled roof, fake arch, exposed post and lintel and green paint. The broad roof and prominent chimney are classic symbols of "home", except the wide chimney is not what it seems to be (actually much smaller) and the sheltering roof (inspired by the Low House) is split down the middle. The classic string course is broken by the windows. The lintel that joins the two halves looks unusually narrow for the weight above. Thus,

Venturi adds the symbolic arch above, which was just a molding(attached decoration), to give more support-except the arch is also broken! Still, the arch helps break the weight of the facade. The tension between the large and small windows on the front facade contributes to the conflict of a broken arch over a rectangular entrance. To add to the confusion, the building is not trying to be quaint or picturesque but has a very tight modern-like surface. To Venturi, these decorative elements and symbols glamorized cheap materials through association with classical forms—all in an attempt to celebrate middle-class lifestyle and values.

On the ground floor plan, a recess leads to the front door which is hidden from view. The plan is based on the symbolic idea of the fireplace as the center of the house. Space feels ambiguous(creates many ways of understanding) as the staircase collides with the fireplace. In the dining area, the ceiling is half-vaulted(recalling the half-circle arch form tacked-on to the facade). The vault seems to just miss a structural column which supports the flat ceiling above. Thus, the vault appears to be resting on a wall of glass doors! The entire design is full of complexity and contradiction. After squeezing past the fireplace on the way upstairs, one arrives in the master bedroom complete with an enormous arched window. Another set of step with extremely high risers lead to … no where-one last architectural ambiguity.

New Words and Expressions:

rectilinear	a.	直线的,沿直线的
shed	n.	棚,小屋
paradox	n.(logic)	自相矛盾的议论
pot	n.	壶,锅
prominent	a.	突起的
split	v.	劈开,切开
quaint	a.	奇特而有趣的
glamorize	v.	使有迷惑力
recess	n.	休会,休息
tack	v.	钉
squeeze	v.	榨,挤

Post-Reading Exercises:

1. Fill in the blanks with the correct answer:

1) Robert Venturi is famous for his book "_____ and Contradiction" (1966) in which he called attention to the importance of Baroque architecture in contradiction to rectilinear Modernism.

    A. Compact   B. Implicitly   C. Simplicity   D. Complexity

2) Venturi argued _____ an architecture that was neither pure nor picturesque,

but which made the most of complexities, contradictions, ambiguities and paradoxes.

A. at B. in C. on D. for

3) Venturi said that architecture did not have to be "heroic and original". Instead it was "OK" to look back upon rich architectural _____ for inspiration and references.

A. style B. history C. form D. color

4) This house _____ his mother allowed Venturi to build his ideas about complexity and contradiction.

A. with B. for C. by D. on

5) The classic string course is broken by the _____. The lintel that joins the two halves looks unusually narrow for the weight above.

A. door B. roof C. windows D. portico

6) To Venturi, these decorative elements and symbols glamorized cheap materials through association with classical forms-all in an attempt to celebrate _____ lifestyle and values.

A. high-class B. gutter C. middle-class D. noble-class

7) The tension between the large and small windows on the front facade _____ to the conflict of a broken arch over a rectangular entrance.

A. reduced B. contributes C. advised D. starts

8) On the ground floor plan a _____ leads to the front door which is hidden from view.

A. recess B. garden C. window D. bedroom

2. Decide whether the following statements are true (T) or false (F) according to the passage given above.

1) In his later book Learning from Las Vegas (1972), Venturi and Denise Scott Brown apposed the idea of the "Complexity and Contradiction".

2) Venturi said that architecture did have to be "heroic and original".

3) Venturi said that it was "OK" to look back upon rich architectural history for inspiration and references.

4) Venturi's mother sits in a chair with a pot of flowers in her arm to help create a sense of scale.

5) The broad roof and prominent chimney are classic symbols of "home", except the wide chimney is not what it seems to be.

6) The vault has a structural column which supports the flat ceiling above.

7) To add to the confusion, the building is not trying to be quaint or picturesque but has a very tight modern-like surface.

8) To Venturi, these decorative elements and symbols glamorized cheap materials through association with classical forms.

UNIT 4   STRUCTURE

Part 1   Warm-up Activities

1. Look and read some examples of construction systems:

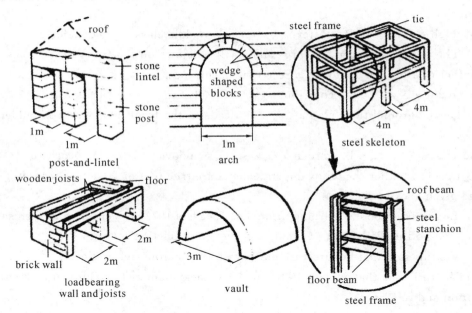

2. Read the following:

The post-and-lintel structure *consists of* three upright posts and two horizontal lintels. Materials used for post-and-lintel structures include stone and timber.

Now make similar statements about the load bearing wall and joist structure.

3. Read the following:

The post-and-lintel structure, in the diagram above, *is composed of* straight members. The vertical and horizontal members which are used to make the structure *are called* posts and lintels respectively, the posts *are spaced at* 1 meter centers. They *are made up* of blocks. Both the posts and the lintels *are made of* stone.

Now write a similar description of the load bearing wall and joist structure.

4. Look at the diagrams in exercise 1 and answer these questions:
a) What do the stanchions carry?
b) What do the floor beams support?
c) What does the steel frame consist of?
d) What is the arch made up of?
e) At what centers are the steel frames spaced?
f) What are the horizontal members which connect steel frames together called?
g) What distance does the vault span?
h) What is the span of the arch?
i) What is the stone in the centre of the arch called?
j) Give some examples of materials used for actuated and framed structures.

Part 2   Controlled Practices

1. Look at the table about the components of a factory. Use the table below to make questions and give the appropriate answers.
Example:
1) What does a roof look like?
2) How many elements does a roof consist of?
3) What is the floor structure made of?
4) What are corrugated sheets made of?

Elements	Compound units	Units	Materials
Roof	roof structure waterproof covering	joists and slabs sheets	timber wood-wool asphalt
Walls	cladding wall structure heavy timber frame light timber frame	corrugated sheets columns and beams stanchions	steel steel
Floors	wearing surface floor structure	tiles panels	vinyl precast concrete
Foundations	column bases pile		concrete wood, precast concrete

2. Look at this section of a factory and label the components using the table in exercise 1:
Example: a) timber, b) joists, c)_____, d)_____, e)_____, f)_____, g)_____, h)_____

3. Now complete this passage:

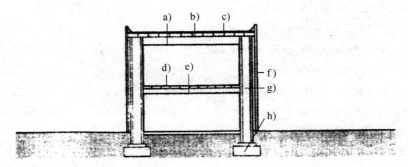

The factory _____ from four elements: the _____, the _____, the _____ and the _____. The roof _____ a waterproof covering, which is made of _____ and a _____, which is made of timber joists and _____ slabs. The walls are constructed from two _____, the wall structure, which consists of _____ and the _____, which is made of _____ sheets. The _____ consists of a wearing surface, which is made of _____ and a floor structure, which is made of _____. The foundations consist of _____.

4. Answer these questions by giving properties of materials.

a) Why is steel used for the frame structure of the factory?

b) Why is asphalt used for the waterproof covering?

c) Why are corrugated steel sheets used for the cladding?

d) Why are vinyl tiles used for the wearing surface?

e) Why is concrete used for the column bases?

Part 3   Further Development

1. Study and read:

The following diagram shows a steel frame structure. The span of the beams measures 9 meters. The distance between the structural frames is about 3 meters.

UNIT 4 STRUCTURE

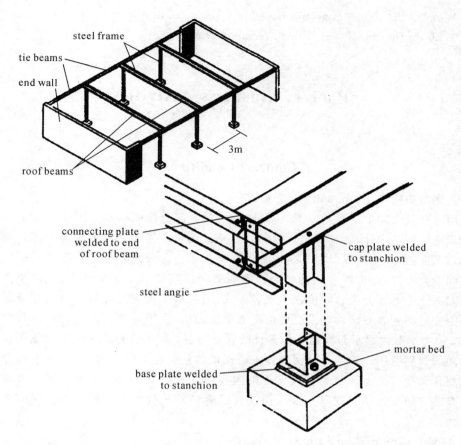

The single-storey structure consists of three frames. These frames are made up of steel stanchions and beams. The frames are placed between end walls and spaced at 3 meter centers. The stanchions carry the beams. These beams support the roof. The roof beams cantilever a short distance beyond the stanchions. This means that they extend over the profiled sheet steel cladding. The cladding can then be placed outside the line of the stanchions.

The beams are bolted to steel stanchion caps. The stanchion caps are welded to the top of each stanchion. The load on each beam is transmitted through these plates to the stanchions.

The upper face of the steel base plates and the ends of the stanchions are machined flat. The bottom of each stanchion is welded to a base plate. Each base plate is fixed to a concrete column base by two holding-down bolts.

Steel angles are fixed across the ends of the beams and built into the brick walls. These angles tie the frames together and also provide a place to fix the top of the cladding.

2. Now answer the following questions:
a) How are the angles fixed to the roof beams?
b) How are the loads of a roof beam transmitted to the column bases?

c) What is the joint between a base plate and a column base filled with?

d) Why do the roof beams cantilever a short distance?

Part 4 Business Activities

Contract Conditions

1. 文化与指南(Culture and Directions)

通用的建筑合同主要包括以下几部分内容:第一部分应明确本合同承包范围及承包方式。例如施工图纸中什么内容不包含在其内,总价承包还是综合单价承包等等。第二部分应明确总工期、节点工期以及对工期的奖惩措施。第三部分应明确工程质量目标。第四部分为合同价款部分,是工程合同的核心。需明确项目整体总价及单体单价,以及合同价款的调整方法。第五部分为合同价款的支付方式,通常按工程进度支付。第六部分应明确施工过程中变更的管理方法。通常变更由施工单位提出,由甲方工程部会同成本管理部门进行审核。第七部分为总承包商对分包单位的管理内容。需明确分包单位的详细工作内容及总承包商的管理费用。第八部分明确业主需提供的建筑施工或安装材料。第九部分明确双方的权利和义务。第十部分为合同文件的内容解释。第十一部分明确合同的争议与违约的解决方法。通常选择工程所在地人民法院或者仲裁机构解决。最后为合同的生效与终止及其他补充条款。

2. 情景会话(Situational Conversations)

(1) Negotiating Contract Conditions and the Price

(A negotiation on Contract Conditions is a meeting held between two parties, the client and the contractor, and the objective is to reach an agreement over issues which are important in both parties' views, or may involve conflict between the parties, or need both parties to work together to achieve their objective. Now the client's consultant engineer and the contractor's project engineer are having a negotiation on both Contract Conditions and the Price.)

(A: The client's consultant engineer; B: The contractor's project engineer)

A: Today we can enter the commercial negotiation stage, in this stage we would like to discuss both contract conditions and the price. We will spend a couple of days on each topic.

B: It is good to have a chance to talk about the contract conditions. We thought it was fixed already.

A: No. It is open for discussion. First, I would like to make it clear that in principle the contract conditions are complied with the FIDIC (Conditions of Contract for

Works of Civil Engineering Construction) document.

B: We quite appreciate that because the FIDIC is relatively fair for both client and contractor.

A: I am sure that you must have read these special conditions carefully and you can present your comments now.

B: Frankly speaking, we do have some different opinions on the conditions if you don't mind. Firstly there is not a clause for the advanced payment for the contractor in the conditions. As usual, the contractor should have around 10% percent of contract value for preparation of works after the contract is signed and then the contractor provides a bank guarantee of same value of the payment.

A: I am afraid that there is no room for discussion in this point. It is the principle of this contract that the payment can be started only from the progress payment. Contractors must prepare the fund for their mobilization cost by themselves but they may include the loan interest into the on cost in their price.

B: Secondly, Clause 10, which is about the performance bond. We think that 15% of the bond is higher than usual practice. We hope it will be modified to 10%.

A: We will consider what you proposed for this Clause.

B: For Clause 41 we have two points of different opinions. One point is that the contractor must commence the Works within 10 days after the signing of the contract. We think it is a little bit too tight.

A: We don't think so. In the definition in Clause 1, it has made clear that this project includes both the permanent works and temporary works. It is quite enough for your people to move in and set up the construction equipment on site to start your temporary works within 10 days. For such a tight construction program we can not allow any delay for commencement.

B: In such a case we withdraw our suggestion for this point. The other point we'd like to propose is that in Clause 41, it only mentions what the contractor should do but does not mention what the Contractor is entitled to do in case the site is not handed over to him on the time agreed by both sides.

A: It will not happen in such a situation. The site will be ready for handing over to the contractor before the contract is signed.

B: Even if it will not happen, but we insist on adding a sentence like this: If the contractor suffers delay or incurs costs for possession of the site or a part of the site, he has the right to obtain an extension of time to complete the works and an amount of cost added to the contract price.

A: We still think it is unnecessary to write down a sentence for something that will never happen.

B: I am afraid I can not agree with you that contract conditions are only written for the situations that are likely to happen. For example, there is a Clause for arbitration. We believe it will not happen as long as both sides strictly follow the all things defined in the contract, but it is still mentioned in the contract.

A: It seems that in this point we can not reach an agreement. I propose to leave it on the final discussion. Now let's have a tea break.

(2) Discussing about Other Clauses

(After the tea break.)

A: Let's go on for Clauses of the Conditions.

B: Concerning Clause 51 variation orders, it is mentioned that the employer shall issue variation orders to the contractor for increasing or deducting a part of the works and contractor shall do any of the works instructed by the engineer. We think it should have a limitation of the variation works.

A: It is also mentioned in the same Clause that the contractor will be paid on basis of the rates agreed in the contract for the works done for the variation orders.

B: Suppose that the total increased or decreased works exceed 30% of the initial contract price, then all resources of material, machines, labor and facilities will be much different from the original estimation, it is sure the surrounding of the contract rates are changed, so the rates for the payment to variation should be modified.

A: It can be sure that the variation orders will not become so big an amount. We have no objection to making a limitation for the variation, say, makes it less than 30% of the total value of the initial contract price, as you insisted. But the Employer should determine the sum paid to the contractor for it.

B: We think the sum should be negotiated by both sides.

A: That's reasonable.

B: There is one thing very important to the progress of the works, for which we must point out here. It is mentioned in the Clause 61 drawing supply that the engineer will supply all drawings in accordance with the necessity of the progress of the works defined in the drawing supply schedule. But what shall we do if the delay of the drawing supply really happens?

A: We will do everything to avoid it.

B: Our proposal is to add a sub-clause like that in case the contractor suffers delay of drawings supply he is entitled to have an extension of time to the contract period and an amount of a cost for compensation to catch up with the progress will be added to the contract price. We think the above two points are very reasonable and should be added in this Clause.

A: We agree with your above two points in principle. But we must add a sentence after the two points that the time extension and cost increase will be determined by the Engineer after due consultation with the employer and the contractor.

B: OK, we agree to your modification to this new sub-clause.

A: It is necessary to have one more sub-clause for this important issue like that the contractor shall give notice to engineer within a reasonable time when he thinks it will cause delay to the works without being supplied with the drawing.

B: As an experienced power station builder we can do that for good will but will not take any responsibility for the delay of drawing supply.

A: Of course not. It just shows the co-operative spirit from the contractor. Are there any more comments to the contract conditions?

B: We would like to discuss something about the Clause 62 claims if you don't mind.

A: Never mind, please go on.

B: We think the procedure of claims is rather complicated. It is mentioned in the Clause that there are 10 steps should be fulfilled for each claim.

A: Since the claim is a very particular matter for implementation of a contract, both sides of the contract, should deal with it carefully.

B: In our opinion, claim events often happen in construction field. In such a size of the project, it is usual that several hundreds of claims may occur. We can not imagine how much time and how many steps we should take for every claim according to this procedure. It will surely waste us a lot of time and energy.

A: But there is nothing special in this contract conditions compared with other contract documents.

B: And also in the Clause we can find that after the event of the claim is finished the engineer is entitled to determine the amount of claim within 135 days. So both of us will take several years to solve all the claims after the works has been completed. Are you sure it is suitable for both sides to do so?

A: Please believe in me that there are no conflicts between this contract conditions and normal international contract documents. There is no doubt that all these steps of the procedure for claim are necessary.

B: It is hard for me to understand your explanation. We think our discussion should be on reasonable basis.

A: Yes, it is being on reasonable basis. Now let's finish the discussion for the contract conditions and turn to the next part of the commercial negotiation. If your comments are not yet expressed totally, please put them in your letters. We may possibly arrange a discussion later. Is this arrangement all right for you?

B: OK. We have to say that it is all right. We will put our comments to the contract conditions in written form and submit to you soon for your consideration.

A: That's fine. So see you the next meeting.

B: See you.

New Words and Expressions:

be complied with	遵循,遵照
appreciate v.	赞赏,欣赏,感谢
on cost	间接成本;间接费用
propose v.	提出(建议、动议等)
commence v.	开始
entitle v.	使有资格,给……定名,给予……权利
incur v.	招致
likely a.	可能的
arbitration n.	仲裁
define v.	确定
claim n.	索赔,主张(权利等)
mobilization n.	运作,使用
modify v.	修改,调整
fulfill v.	完成
implementation n.	执行

Notes to the Text:

1. FIDIC是"国际咨询工程师联合会"法文名称 Fédération Internationale Des Ingénieurs-Conseils 的前5个字母,其英文名称是 International Federation of Consulting Engineers。FIDIC 于1913年由欧洲5国独立的咨询工程师协会在比利时成立,现总部在瑞士洛桑。这是一个国际性的非官方组织。中国工程咨询协会——CNAEC(China National Association of Engineering Consultants)于1996年成为其成员。FIDIC 编制的《土木工程施工合同条件(红皮书)》、《电气与机械工程合同条件(黄皮书)》、《工程总承包合同条件(桔黄皮书)》被世界银行、亚洲开发银行等国际和区域发展援助金融机构作为实施项目的合同和协议范本。

2. contract conditions and the price:合同条款和价格。

3. special conditions:(合同的)特殊条款。general conditions 为"一般条款"。

4. bank guarantee:银行保函,银行保证书。

5. progress payment:由业主支付给承包商的工程进度款。

6. performance bond:履约担保,履约保函。

7. in clause 41, it only mentions what the contractor should do but does not mention what the contractor is entitled to do in case the site is not handed over to him on the time agreed by both sides. 合同第41条仅仅提及承包商应该做什么,但未提及(业主)如若不能按合同中双方约定的时间将施工现场移交给承包商时后者应得的权利。

8. he has the right to obtain an extension of time to complete the Works and an amount of cost added to the contract price:他(承包商)有权获得工程完工延期的许可以及(相应的)合同总价的费用的补偿。

9. variation orders:(施工项目)变更指令。指在实际施工过程中由业主代表和工程师所签发的和原施工清单和图纸中施工项目所不同的新的施工内容。业主工程师在下达这些指令时必须给予承包商合理的额外的施工时间和费用,这些变更一般不得超过合同总量的一定比例,如果变更指令项目过多将会影响承包商按合同要求按期完工,尤其是关键工序的重大变更,则必须给承包商批准相应的延期完工的许可。

10. give notice to:给……下达书面通知。

11. We think our discussion should be on reasonable basis:我方认为讨论应该建立在合理的基础上。

Post-Reading Exercises:

Translate the following sentences into Chinese:

1) First, I would like to make it clear that in principle the contract conditions are complied with the FIDIC (Conditions of Contract for Works of Civil Engineering Construction) document.

2) Our proposal is to add a sub-clause like that in case the contractor suffers delay of drawings supply he is entitled to have an extension of time to the contract period and an amount of a cost for compensation to catch up with the progress will be added to the contract price.

3) It is necessary to have one more sub-clause for this important issue like that the contractor shall give notice to engineer within a reasonable time when he thinks it will cause delay to the works without being supplied with the drawing.

Part 5   Extensive Reading

(A) Structural Design

A structure is the part of a building that carries its weight, and for at least half the world's civil engineers, structures are most of civil engineering. A building is a structure with a roof and much of civil engineering structural design is the design of building structures. The building as a whole is designed by an architect.

The structural design itself includes two different tasks, the design of the structure, in which the sizes and locations of the main members are settled, and analysis of this structure by mathematical or graphical methods or both, to work out how the loads pass through the structure with the particular members chosen.

For the typical multi-storey structure in a city, whether it is to be used for offices or dwellings, the most important member which the engineer designs is the floor. There are two main types of reinforced concrete floors, the solid floor and the hollow tiled(a ribbed) floor. In the ribbed floor, part of the lower half of the slab is hollow which has a great

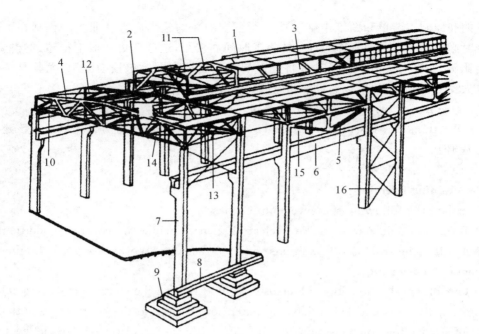

1. roof boarding 2. skylight truss 3. skylight cladding 4. truss 5. bracket 6. crane beam 7. pillar 8. foundation beam 9. foundation 10. continues beam 11. skylight brace 12. horizontal brace of upper truss 13. vertical brace of truss 14. horizontal brace of lower truss 15. vertical brace of lower truss 16. brace between pillars

advantage because this concrete would not strengthen the floor but would be heavy. Ribbed floors are therefore lighter than solid floors; but it is more difficult to cast them with holes through them unless these holes are carefully planned before hand.

Among the most interesting structures at the moment are suspended structures. In all this structures, the columns are made fewer and larger so as to reduce the bulking effects on them and to increase their effective length. In two that were recently built in London, there is only one column, in the center of the building, and this is a hollow concrete tower some 12m square which carries the lifts, stairs, ducts, pipes, and cables within it or attached to its wall. The tower may be called the core of the building and on its top is a bridge overhanging in all directions, from which high tensile steel bars drop to carry the floors below.

These bars are very thin and can be hidden in a door frame or window frame so that for such a building, there need be no noticeable obstruction to sight or horizontal movement in any direction outwards from the core.

But this is only the beginning of suspended construction if it is successful and if the world's large cities continue to become more crowded, the idea will grow, and the 60-storey skyscrapers of New York will be tiny compared with the vast 300-storey structures of the world's future cities.

It seems possible and even likely that the whole city may be one or a few of these vast buildings, carried on pairs of towers 1,000m high joined by light weight bridge structures, possibly suspension bridges.

Notes to the Text:

1. the ribbed floor: ribbed,过去分词作定语,意义相当于 the floor which was ribbed,译为"肋形地板"。

2. there need be no noticeable obstruction to sight...: "对视线不能有什么明显障碍", movement 在此句中指人的走(活)动。

3. ... and this is a hollow concrete tower ... or attached to its wall: 这是一座约12平方米的空心混凝土塔,它支撑着其内部或设在壁上的电梯、楼梯、风道、管道以及电缆。

New Words and Expressions:

dwell	v.	居住
dwellings	n.	住宅
location	n.	位置,场所
graphical	a.	图的,图解的
typical	a.	典型的,代表性的
solid floor	n.	实心地板
hollow	a.	空的,凹的
rib	n./v.	肋,加肋
slab	n.	板,石板
strengthen	n.	加强,加固
beforehand	ad.	预先
stair	n.	楼梯,阶梯
attach	n.	缚,系,附加
core	n.	核心
overhang	v.	伸出,悬垂
obstruction	n.	障碍,阻塞
lightweight	a.	轻质的
buckling effect	n.	压屈效应

Post-Reading Exercises:

1. Translate the following expressions:
1) carry a structure's weight
2) be designed by an architect
3) the analysis of this structure by graphical methods
4) the loads pass through the structure
5) the typical multi-storey structure
6) the solid floor and the hollow ribbed floor

7) cast them with holes through them

8) suspended structure

8) reduce the bucking effects

10) carry the lift, stair, duct and cable

11) 房屋建筑结构的设计

12) 用作住宅的办公楼

13) 在伦敦建成的两幢建筑物

14) 面积约12米见方的空心混凝土塔架

15) 建筑物的核心结构

16) 从桥上伸下高强钢筋悬吊

17) 隐藏在门框或窗框里

18) 对视线没有明显的障碍

19) 设计思想的发展

20) 用轻型桥梁结构连通

2. Fill in the missing words:

Among the most interesting structures at the moment are 1)_____. In all this structures, 2) are made fewer and larger so as to reduce 3)_____ on them and to increase their effective length. In two that were recently built in London, there is only 4)_____, in the center of the building, and this is 5)_____ some 12m square which carries the lifts, stairs, ducts, pipes, and cables within it or attached to its wall. The tower may be called 6)_____ and on its top is 7)_____ overhanging in all directions, from which high tensile 8)_____ drop to carry the floors below.

These bars are very 9)_____ and can be 10)_____ in a door frame or window frame so that for such a building there need be no noticeable obstruction to sight or horizontal movement in any direction outwards from the core.

3. Put the following sentences into English:

1) 结构物是建筑物的一部分,它支撑建筑物的重量。

2) 弄清楚所选特定的结构物中荷载是如何传递的。

3) 楼板是工程师设计的最主要的构件。

4) 肋形楼板比实心楼板要轻得多。

5) 悬挂结构柱子数量较少而尺寸较大,以降低对其压屈作用,增加其有效长度。

(B) Structure Members of A Building

The structure of all buildings is made up of various combinations of structural elements such as beams, arches trusses and columns. They are assembled so that each may perform its function in the structure.

Beams are the horizontal members used to support vertical applied loads across an opening. In a more general sense, they are structural elements that external loads tend to bend or curve. There are many ways in which beams may be supported.

The beam is called a simply supported or simple beam. It has supports near its ends, which restrain it, only against vertical movement. The ends of the beam are free to rotate. The beam is a cantilever. It has only one support, which restrains it from rotating or moving horizontally or vertically at that end. When a beam extends over several supports, it is called a continuous beam.

An arch is a curved beam, which radius of curvature is very large relative to the depth of the section. It differs from a straight beam in the following ways: (1) Loads induce both bending and direct compressive stresses in an arch. (2) Arch reactions have horizontal components even though loads are all vertical. (3) Deflections have horizontal as well as vertical components. Arches may be built with fixed ends as can straight beams, or with hinges at the supports. They may also be built with a hinge at the crown.

The vertical members of a structural frame are called columns. The column is an essentially compression member. The manner in which a column tends to fail and the amount of the load causes failure depends on the material of which the column is made, and the slenderness of the column. The first point is obvious. For example, other conditions(size, height) being identical, a steel column is capable of supporting a much greater load than a timber column. When a column is very long and slender, failure will occur due to buckling at a much lower load than would cause failure in a short column of equal cross-section. The shape of a column is also very important. For example, a sheet of cardboard has practically no as strength as a column, but if bent to form an angle section or other shapes, it is capable of supporting a load.

A truss is a framed structure consisting of a group of triangles arranged in a single plane in such a manner that loads applied at the points of intersection of the members will cause only direct stresses (tension or compression) in the members. Since a truss is composed of a group of triangles, it is possible to arrange innumerable types. But certain types have proved to be more satisfactory than others, and each of these has its special uses.

Notes to the Text:

1. Arches may be built... supports: 拱可以像梁那样,做成两端固定,也可以把两支点做成铰接的。

2. When a column is long and slender,... failure in a short column of equal cross-section: 当一根细长的柱子压曲而发生断裂时,与一根横截面相同但较短的柱子相比,所承受的荷载要低得多。

New Words and Expressions:

element   n.   要素,(建)构件
opening   n.   空间,孔洞,开口
truss   n.   桁架,构架

combinations	n.	联合,组合
horizontal	a.	地平的,水平的
vertical	a.	垂直的,直立的
curve	v.	(使)弯曲
restrain	v.	制止,约束
rotary	a.	转动的
radius	n.	半径
curvature	n.	曲率
section	n.	截面,部面
induce	v.	引诱,诱导,引起
compressive	a.	有压力的,压缩的
compressive stress		压应力
deflection	n.	挠度,曲,偏转
hinge	n.	活关节,铰接,铰链
crown	n.	冠,拱顶
slenderness	n.	细长
buckling	a.	翘曲,弯曲
cross-section	n.	横截面,横断面
intersection	n.	横断,交叉,交叉点
tension	n.	拉,拉力
innumerable	a.	无数的,数不清的
frame structure		构架结构,系统结构
restrain from		制止,约束
a point of intersection		交点
radius of curvature		曲线半径

Post-Reading Exercises:

1. Multiple Choices:

1) Beams are the horizontal members used to _____ loads across an opening.

    A. apply   B. arrange   C. carry   D. pull

2) A simple beam has supports near its ends. It cannot _____ freely.

    A. move horizontally   B. move vertically

    C. rotate   D. turn around

3) An arch is a curved beam. Its radius of curvature is very large _____ to the depth of the section.

    A. absolute   B. measured

    C. in regard   D. in relation

4) A steel column is able to support a much greater load than a column made of _____ .

A. plastic　　　　B. wood　　　　C. concrete　　　　D. glass

5) The manner in which a column tends to fail and the amount of the load causes failure depends on the material of which the column is made, and the of the column.

A. shape　　　　B. thinness　　　　C. height　　　　D. weight

6) All triangles of a truss are arranged in the same plane, so that loads applied at the points of intersection of the members will cause _____ in the members.

A. compression　　　　　　　　　　B. tension

C. tension and compression　　　　D. direct stresses

7) For example, _____ has practically no the strength as a column, but if bent to form an angle section or other shag it is capable of supporting a load.

A. a plastic bag　　B. a tissue　　　C. cloth　　　　D. paper

8) Due to the fact that a truss is composed of a group of triangles, it is possible to arrange _____ types.

A. several　　　　B. some　　　　C. countless　　　　D. endless

2. Answer the following questions:

1) Give examples of structural elements in a building.

2) In what ways can a beam be supported?

3) What is a cantilever? Explain the technical term.

4) What are main differences between an arch and a beam?

5) Why and in what way is the shape of a column important?

6) What is a truss?

7) Can a truss consist of several squares and circles arranged in a single plane?

8) Which kinds of stresses do loads cause that are applied at the points of intersection of a common truss?

UNIT 5   FOUNDATION

Part 1   Warm-up Activities

1. Look at these pictures, read aloud the English labels and translate them into Chinese.

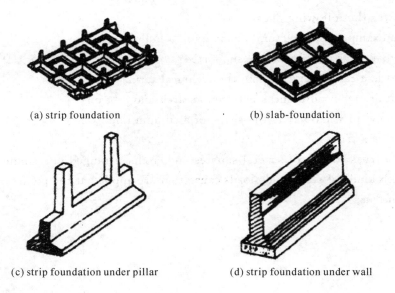

(a) strip foundation   (b) slab-foundation

(c) strip foundation under pillar   (d) strip foundation under wall

While the term foundation is sometimes used to mean the ground supporting a building or civil engineering structure, it generally refers to the part of a structure which transfers the "load" to the ground. Although it does not show in the finished structure, the foundation is one of the most vital parts of the work, and may cost more than the visible structure itself. In order to provide sound foundation, the engineer designing them must know what loads are to be supported and what the properties of the ground are below.

2. Look at these diagrams, read out the labels and translate them into Chinese:
Concrete Foundation
1) rising wall
2) connecting rods, starter bars

3) supporting formwork
4) horizontal construction joint
5) natural soil
6) in-situ concrete
7) bedding, blinding bed
8) angle of load distribution
9) angle of natural slope

Reinforced Concrete Foundation
1) rising pier
2) reinforcement bar
3) reinforcement stirrup of binder
4) construction joint
5) bent reinforcement bars
6) blinding concrete
7) bearing soil
8) poor bearing stratum
9) straight reinforcement bars

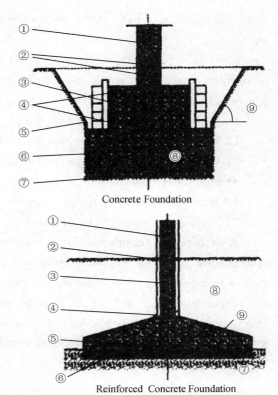

Concrete Foundation

Reinforced Concrete Foundation

Part 2　Controlled Practices

1. Read aloud the following passages about foundation loads and learn the new technical vocabularies.

Foundation loads: A foundation supports different kinds of loads, such as:

Dead Load. Dead load is a fixed position gravity service load, so called because it acts continuously toward the earth when the structure is in service. The weight of the structure is considered dead load, as well as attachments to the structure such as pipes, electrical conduit, air-conditioning and heating ducts, lighting fixtures floor covering, roof covering, and suspended ceilings; that is, all items that remain throughout the life of the structure.

Live Load. Gravity loads acting when the structure is in service, but varying in magnitude and location are termed live loads. Examples of live loads are human occupants, furniture, movable equipment, vehicles, and stored goods. Some live loads may be practically permanent, others may be highly transient. Because of the unknown nature of the magnitude, location and density of live load items, realistic magnitudes and the positions of such loads are very difficult to determine.

Snow Load. The live loading for which roofs are designed is either totally or primarily a snow load. Since snow has a variable specific gravity, even if one knows the depth of snow For which design is to be made, the load per unit area of a roof is at best only a guess.

Wind Load. All structures are subject to wind load, but they are usually only those more than three or four stores high, other than long bridges? For which special consideration of wind is required.

2. Answer the following questions:

1) What is dead load? Use your own words to describe it.

2) What does live load mainly consist of?

3) How does one calculate snow load?

4) Which buildings are mainly exposed to wind load?

3. Translate the following passage:

A satisfactory foundation for a building must meet 3 general requirements: 1) the foundation, including the underlying soil and rock, must be safe against a structural failure that could result in collapse. 2) During the life of the building, the foundation must not settle in such a way as to damage the structure or impair its function. 3) The foundation must be feasible both technically and economically, and practical to build without adverse effects to surrounding property.

Part 3 Further Development

1. Read the following English and Chinese expressions for ground foundation and try to memorize them.

foundation	地基
earth(ground) endurance	地耐力
geological prospecting	地质钻探
earthquake magnitude	地震震级
sand foundation	砂基础
pile foundation	桩基础
rubble foundation	毛石基础
reinforced concrete	钢筋混凝土
brick foundation	砖基础
deep foundation	深基础
shallow foundation	浅基础
quick sand	流沙
ground beam	地梁
foundation beam	基础梁
damp-poof course	防潮层

UNIT 5 FOUNDATION

footing course	承台底层,基(础)层
raft foundation	木筏式基础
rigid foundation	刚性基础
shell foundation	壳体基础
plain concrete saddle	无筋混凝土(素土)垫座
sand bedding course	砂垫层
machine-rolled foundation	机械碾压地基
slope of foundation pit	基坑边坡
drilling-poured pile	钻孔灌注扎
explosion-poured pile	爆破灌注桩
steel sheet pile	钢板桩
sandy soil	砂土
expanded clay	膨胀黏土

2. Make sentences with the words given below, following the patterns provided:

1) the purpose of a foundation, be, carry, the load of a structure, and, spread, it, over a greater area(The purpose of... is...)

2) the bearing capacity of a soil, mean, the maximum load, per unit area, which, the ground, safely, support(... means...)

3) as, the nature of the soil, often, vary, considerably, on the same construction site, the capacity of the soil, support loads, also, vary(... often vary considerably...)

4) it, be not, always, possible, provide, a uniform size of foundation, for the entire structure(It is not always possible to provide...)

5) a foundation, normally, consist of, either plain, or, reinforced, concrete, which, should, lay, sufficiently, below, the ground frost level, avoid, the possible danger, of, frozen soil, lift it(... normally consist of...)

6) soft sports, be, usually fill with, consolidated hardcore, or, a weak concrete(... be normally supported on...)

7) isolated columns, or, stanchions, be, normally, support, on, square concrete foundation bases(... be usually filled with...)

8) such columns, space, at close intervals(... be spaced at... intervals)

9) it, be, more practical, provide, a continuous concrete strip foundation, carry, a complete row(It is of ten more practical to...)

10) a raft foundation, be, often, recommend, support, normal buildings(... be often recommended...)

3. Translate the following sentences into Chinese:

1) The size and type of foundation depends on the nature of the ground and the load carried by it.

2) As the nature of the soil often varies considerably on the same construction site, the capacity of the soil to support loads also varies.

3) Where such columns are spaced at close intervals, it is often more practical to provide a continuous concrete strip foundation to carry a complete row, as is done for load bearing walls.

4) Where the bearing capacity is particularly poor or the quality of the soil varies considerably, a raft foundation is often recommended to support normal buildings.

5) When a foundation is greater depth than normal to laid in poor soil, it is necessary to excavate to a reach a solid base.

Part 4　Business Activities

Price Negotiation

1. 文化与指南(Culture and Directions)

工程承包的价格谈判,是企业取得理想经济效益的关键一环。价格谈判中应注意以下几点:

首先,要掌握多方面知识,如建筑施工、技术材料、各种法规、内外部环境等,对对方提出的质疑做到心中有数,能给出满意的解释,做到有备而来,才能掌握谈判的主动权。

其次,熟悉谈判技巧,掌握谈判策略。先让对方表明其要求,自己可以在非关键问题上作适当让步,以给予对方心理上的平衡,但让步幅度不要太大太快,一般来讲,让步都要从对方获得相应的益处。在谈判桌上,双方各自代表本公司的利益。如果感觉有必要说"不",就应该勇敢地提出来,只要你说得有道理,相信对方会接受你说的"不"。

再者,说服对方,自己要心中有数。在谈判的开始阶段,一般尽量先讨论那些容易解决的问题,不要一开始就使气氛紧张,这样不利于解决问题。用互惠互利的条件说服对方。在谈判中,强调对方许多有利的因素,激发对方在自身利益认同的基础上接纳你的意见和建议。

2. 情景会话(Situational Conversation)

A business negotiation is holding on. The purpose of a business negotiation is to reach agreement, not to score points in argument. However, since the offer is in a higher point, and the counter-offer is in a lower point, bargaining is an unavoidable thing in BN. And actually, it is usually the core of a BN. In most circumstances, quick settlements should be avoided. The outcomes often favor the more experienced negotiator.

(A: Consultant Engineer of South Power station; B: Representative of the contractor)

B: Good morning! Nice to meet you again!

A: Me too. Today we are going to discuss the tender price. Generally speaking, your price is very high compared with the other bidders.

B: The price of this project is worked out on the basis of our experience for power plant construction. We have reduced our profit to a very low level. Considering the benefit of the employer, we believe that the timely completion of the works with good quality at a reasonable price is more significant than the delay of power producing. Nobody would expect a problematic situation just for a little lower price.

A: Why can't you do a good job with a competitive price?

B: There is an old Chinese saying that you can not let a horse run fast without feeding enough grass.

A: Now let's have a review of the tender price in your submission. A breakdown of your prices will give us explanation whether your price is reasonable or not.

B: We are pleased to do so. The total price consists 9 items. Item 1 is the labor cost. The total labor cost is USD 6.39 million, around 10% of the total price.

A: How many man-hours do you estimate to spend for this project?

B: About 8.9 million man-hours in minimum. So the labor cost is only USD 5 per day which is rather low. Nearly all labor force will be employed locally. The Item 2 is the staff cost. It is around USD 1.9 million on total.

A: From our calculation, your staff cost is rather high. The monthly salary is about USD 1500, it is very high compared with the average staff salary in your Country.

B: This amount is not the actual salary of the staff. In fact only 65% becomes their income, the other 35% is the cost for transportation, telecommunication, tax, insurance, and so on.

A: We would like to know the content of your staff cost.

B: OK. We will submit a list of the staff cost after the meeting. Item 3 costs is the construction equipment cost. The total of it is USD 7.196 million.

A: It is a very high cost. Can you tell us what percentage you calculate for the depreciation of the equipment?

B: There are several depreciation periods we used for the equipment. For example, for large and durable equipment, it is 15 years. Some equipments are 10 years and some easily consumable machines are 3 years. For such a large project with a tight program, many high efficient equipments such as concrete pumps, high capacity concrete batching plants, concrete mixing trucks and many kinds of cranes would be employed which are shown in our Construction Equipment List.

A: We understand that you should mobilize enough equipment for the construction, but in order to control the machinery cost, you'd better prepare a Schedule of Equipment Mobilization which is closely matched with Construction Schedule to avoid any idling of equipment on site.

B: OK. We will check the equipment cost and see if there are any costs that can be deducted. The Item 4 cost is the material cost which is the largest one and occupy

55% of the total cost.

A: From the breakdown we see the cement cost USD 4.97 million. How much is the purchasing price for each tone of cement?

B: From our investigation, the cement is USD 70 per tone including the cost of material, transportation and other costs. We have already underestimated this cost. The risk of inflation in next three years has not been considered yet.

A: It seems that 71,000 ton of cement for around 220,000m^3 concrete is higher than usual.

B: We can not judge this item by average calculation. There are several grades of concrete for different parts of the works and the mixture of each grade of concrete is different. And we should also think of the consumable loss during delivery and operation.

A: We had better have a detailed breakdown for concrete, reinforcement and formwork in later stage. Now let's go on the total price breakdown.

B: Item 5 cost is for technology transfer and training of local workers and staff. For this item, we take 1.5% from the total price. Item 6 is the cost for the implementation of quality assurance.

A: I hope you can explain the difference between these two items of costs.

B: Item 5 is the cost only for the local people. We will train the local staff how to make arrangement for a project in such a scale, how to schedule of the works program and how to operate the construction plants and the testing instruments. Item 6 is the cost for all people to take necessary measures for a comprehensive quality assurance operation. For example, the investigation for material suppliers and the testing of materials will be carried out more carefully than usual.

A: You still can not convince me adequately for why it is necessary to have additional cost. Further explanations for these two items are still necessary.

B: That's all right. Item 7 is the cost for tax provision which include the business tax and profit tax regulated by the local authority. Item 8 is the camp cost.

A: Concerning the Item 8 cost, it is very high for labor and staff camp. Your budget for the temporary house is USD 60/m^2. It is unnecessary because the local workers can use the house made by a local plant which costs USD 35/m^2 only.

B: But the cost for the camp is not only for house itself, but also cover the other facilities such as fire extinguishing, water and electrical supply, drainage as well as the maintenance for the period of 3 years. And also the staff accommodation cost is more than USD 65/m^2.

A: So we need a detailed cost breakdown for the camp which is divided into two parts: for staff and for labor. I do not think that the cost for staff living will exceed USD 60/m^2 and cost for labor USD 40/m^2. In this case, there is a big amount of price which can be deducted.

B: We will reconsider it and work out a new budget for the camp. Item 9 cost is the major subcontractors cost. They are sub-contractors for the prestressing works, cladding and roofing, HVAC (Heating, Ventilation and Air Condition) works.

A: We can not accept the cost for these special works submitted to us by such several lump sum prices. You are requested to provide us the detailed analysis the same as you did on concrete, formworks and others in the Bill of the Quantities.

C: I see. We misunderstood this point. Actually there is a draft of the detailed subcontracts' cost available. We will resubmit this part of the budget as soon as possible.

New Words and Expressions:

tender price	标价
bidders *n.*	投标者
work out	计算出
on the basis of	根据
power plant construction	修建电厂
timely completion	按时竣工
submission *n.*	呈交
breakdown of prices	价格分析
in minimum	最少
staff cost	人工费
depreciation of the equipment	设备耗费
durable equipment	耐用设备
easily consumable machine	易耗损机械
concrete pump	混凝土泵
concrete batching plant	混凝土搅拌站
concrete mixing truck	混凝土搅拌车
closely matched with	与……匹配
idling of equipment	设备闲置
total price breakdown	总价格分析
technology transfer	技术转让
comprehensive quality	完备的质量
additional cost	附加税
business tax	营业税
profit tax	利润税
camp cost	营地费用
fire extinguishing	消防
subcontractors cost	分包商的费用

Notes to the Text:

1. Schedule of Equipment Mobilization：机械设备进场计划表。

2. take necessary measures for：采取必要的措施。

3. There is an old Chinese saying that you can not let a horse run fast without feeding enough grass. 中国有句俗话：草不足，马不快。

4. The total price consists 9 items. 总价包括 9 项。

5. The total labor cost is USD 6.39 million, about 10% of the total price. 总人工费为 639 万美元，约为总价的 10%。

6. This amount is not the actual salary of the staff. In fact only 65% becomes their income, the other 35% is the cost for transportation, telecommunication, tax, insurance, and so on. 这个数目不是管理人员的实际工资。事实上只有 65% 是他们的收入，其余 35% 为交通费、通讯费、税收、保险、现场外管理费等等。

7. Can you tell us what percentage you calculate for the depreciation of the equipment? 你们能告诉我们采用多少的机械折旧率来计算吗？

8. ... many high efficient equipments such as concrete pumps, high capacity concrete batching plants, concrete mixing trucks and many kinds of cranes would be employed which are shown in our Construction Equipment List. 对这样工期紧张的大型工程，清单中许多必须采用的高效率的施工设备，如混凝土泵、大容量的混凝土搅拌站、混凝土搅拌车和许多种类的吊车等等都需要雇用。

9. ... you'd better prepare a Schedule of Equipment Mobilization which is closely matched with Construction Schedule to avoid any idling of equipment on site. 你们最好准备一份与施工计划表相匹配的机械设备进场计划表来避免现场机械设备的闲置。

10. And also we should think of the consumable loss during delivery and operation. 我们还应当考虑混凝土运输和浇筑时的损耗。

11. ... take necessary measures for a comprehensive quality assurance operation. For example, the investigation for material suppliers and the testing of materials will be carried out more carefully than usual. 为了综合质量保证措施的运行而采取必要的措施……例如，对材料供应商的调研、材料的测试将比通常进行得更仔细。

12. Item 7 is the cost for tax provision which include the business tax and profit tax regulated by the local authority. 第 7 项为税金，包括当地政府规定的营业税和利润税。

13. But the cost for the camp is not only for house itself, but also cover the other facilities such as fire extinguishing, water and electrical supply, drainage as well as the maintenance for the period of 3 years. 但营地费用不仅仅包括房屋本身，而且包括其他设施，如消防、水电供应、排水以及 3 年的设备维护费用。

14. Item 9 cost is the major subcontractors cost. They are sub-contractors for the prestressing works, cladding and roofing, HVAC (Heating, Ventilation and Air Condition) works. 第 9 项费用为主要分包商的费用。他们是预应力工程、墙围护和屋面工程、暖通空调工程的分包商。

Post-Reading Exercises:

Translate the following sentences into Chinese:

1) The price of this project is worked out on the basis of our experience for power plant construction. We have reduced our profit to a very low level. Considering the benefit of the employer, we believe that the timely completion of the works with good quality at a reasonable price is more significant than the delay of power producing. Nobody would expect a problematic situation just for a little lower price.

2) There are several depreciation periods we used for the equipment. For example, for large and durable equipment, it is 15 years. Some equipment is 10 years and some easily consumable machines are 3 years. For such a large project with a tight program, many high efficient equipments such as concrete pumps, high capacity concrete batching plants, concrete mixing trucks and many kinds of cranes would be employed which are shown in our Construction Equipment List.

3) Item 5 is the cost only for the local people. We will train the local staff how to make arrangement for a project in such a scale, how to schedule of the works program and how to operate the construction plants and the testing instruments. Item 6 is the cost for all people to take necessary measures for a comprehensive quality assurance operation. For example, the investigation for material suppliers and the testing of materials will be carried out more carefully than usual.

Part 5 Extensive Reading

(A) Foundations

During the design development phase you should be considering the various types of structural systems that might be used for your building. The system chosen will undoubtedly affect the layout and appearance of the building. The selection of a particular structural system should be finalized just prior to beginning the construction documents.

Most architectural structural systems employ four basic materials: wood, concrete, masonry and steel. They are often used in combination with one another. Glass, plastics, ceramic and other materials are not usually considered to be "structural". The basic parts of the structural system are the foundation, floors, walls and roof.

The function of the foundation is to prevent the building from moving in any direction. A foundation should prevent settling, heaving (upward movement), or lateral shifting. The weight of the building tends to push it downward through the earth. Wind tends to move a building laterally or lift it during severe storms. Earthquakes also typically cause lateral movement. Freezing and thawing of ground moisture and the expansion and contraction of soil that is sensitive to moisture changes may move a building vertically. Also, shifts of the earth may move the building in any direction. The shape and material of the foundation are dictated by soil conditions, economics and occasionally esthetic considerations. The foundation types available are continuous spread footings, pad footings, piling and piers. The continuous spread footings spread the weight of a building over a broad enough area so that the soil is not penetrated by the foundation, allowing moving downward. If some areas of the soil are softer than others, the foundation should still span over these areas without sinking.

Forces tending to move the building upward are common. Every winter, in all but very warm days, the moisture in the ground freezes. As the moisture in the earth freezes, it expands. The expanding earth is strong enough to lift entire buildings. This lifting and lowering(upon melting of the frozen moisture) of buildings or their parts causes cracking of walls and floors because the movement is rarely evenly distributed.

To avoid this type of movement, you must prevent freezing moisture from occurring under the foundation of any building. Since it is impossible to eliminate all of the moisture from the earth, the solution is to place the footings below the frost line. To allow for a rare winter that exceeds historical data, it is necessary to place the bottom of the foundation at least 12 inches below the known frost depth.

The foundation must resist sidewise pressure when it serves as a basement wall. Resistance to sliding of the foundation wall off the footing can be provided by the floor slab, by beating a "key" between the wall and footing, and by reinforcing that interconnects the footing to the wall. Reinforcing rods placed between the footing and the wall also helps prevent the wall from tipping over.

The thickness of the foundation wall should be designed to resist the forces already mentioned. Residential foundations are usually a minimum of 12 inches thick if they are concrete block, or 10 inches if they are concrete. A "rule of thumb" for the size of a footing is that it should be at least twice as wide as the foundation wall and equal to it in thickness. Actual size should be determined by calculations accounting for the strength of the soil and the weight of the building.

Any support columns used in a building typically require a pad footing. Since a column or post transfers a concentrated load to the soil, the final size of the pad must be calculated based on the load applied and the strength of the soil.

Piling is used when the soil near the surface of the site is not structurally sound. In areas near the ocean or other water, the ground may be too weak to support the weight of a building. In areas where expansive soil changes volume with changes in moisture, piling may be desirable to bypass the layer of earth to reach more stable material. A grid pattern of piling will result in a network of sound supports that can be spanned by beams to support the building weight over the undesirable soil.

Wood, steel and round concrete piers are essentially short piling: They are appropriate when great depth is not needed to achieve sound support. If a spread concrete footing(pad) is used, masonry may be added to the list of pier materials.

Notes to the Text:

1. the construction documents:建筑施工说明文件。
2. ceramic tile:瓷砖。
3. continuous spread footing, pad footings, piling and piers:连续扩展式基础,独立式基础,桩和砖墩基础。
4. the soil is not penetrated by the foundation, allowing it to move downward:尽管土壤会让基础向下沉降,但不会被其穿透。
5. the frost line:冰冻线。
6. To allow for a rare winter that exceeds historical data:为了防备罕见的破历史纪录的寒冬(allow for:考虑到,估计)。
7. by the floor slab, by locating a "key" between the wall and footing, and by reinforcing that interconnects the footing to the wall:靠地面的水泥板,靠在墙体和基础之间设置的"插销",以及靠连接基础与墙体的钢筋。
8. tipping over:倾覆,翻倒。
9. A "rule of thumb":凭经验估算(行事)。
10. in area where expansive soil change volume with changes in moisture, piling may be desirable to bypass the layer to reach more stable material:在随着湿度的改变而膨胀性土壤也随之改变体积的地区,最好采取打桩的办法以绕过这种土层而深入到更稳定的地层中去。
11. A grid pattern of piling will result in a network of sound supports that can be spanned by beams to support the building weight over the undesirable soil:网格型桩将会筑起一道牢固的支撑网络,上面可以架梁,以支撑设置在不合乎要求的土壤上面的建筑物重量。

New Words and Expressions:

layout	v.	安排,布局
structural	a.	结构上的
finalized	v.	最后定下来,定稿
prior	a.	优先的
architectural	a.	建筑上的,建筑学的
combination	n.	结合,联合
thawing	v.	缓和
moisture	n.	湿度,湿气
expansion	n.	扩充,伸展
sensitive	a.	敏感,灵敏
esthetic	a.	艺术的
cracking	n.	裂缝,噼啪声

distributed	v.	分配
eliminate	v.	消除
frost	n.	严寒,冰冻
side wise	a.	侧身而行
pressure	n.	压力
sliding	n.	滑行
reinforce	v.	加强,加固
rods	n.	杆,棒
calculations	n.	估计,策划
pad	n.	垫料,护垫
piling	n.	桩,堆积
grid	n.	网格
span	n.	横跨

Post-Reading Exercises:

1. Multiple choices:

1) All the preparation work should be _____ before the construction of a building starts.

A. begun B. done C. carried D. finalized

2) During the construction process, materials such as wood, concrete, masonry and steel are often used in _____ with one another.

A. accordance B. agreement C. combination D. coordination

3) In areas where earthquakes may take place, the foundation of a building should be strong enough to prevent _____ movement.

A. continuous B. downward C. lateral D. upward

4) Yesterday's rain _____ right through my coat as I hurried home without an umbrella.

A. cut B. penetrated C. broke D. wetted

5) If a downward movement is not evenly distributed, the lowering of a building may result in _____ of walls and floors.

A. cracking B. splitting C. settling D. heaving

6) A number of boats _____ over as the result of the huge storm sweeping the coast.

A. knocked B. tipped C. bumped D. dipped

7) I never taste what I am cooking—I just do it by _____.

A. chance B. rule of thumb C. luck D. experiment

8) The strength of a country is determined by calculations _____ for its scientific and technological development, financial capabilities, military strength, etc.

A. accounting B. allowing C. admitting D. asking

9) In areas where the soil is too weak to support a building, it is _____ to use piling.

   A. agreeable   B. convenient   C. desirable   D. capable

10) The new bridge under construction across the river will be _____ by steel beams.

   A. spanned   B. spread   C. set   D. swept

2. Answer the following questions:

1) What can be said about the selection of structural systems of a building?

2) What elements prevent the building from sinking into the ground?

3) By what are the shape and material of the foundation determined?

4) What are the directions of the forces that must be resisted by the foundation wall?

5) If a foundation wall is 14 inches thick, what should be the minimum size of the footing(Use the "rule of thumb")?

6) What is the function of a grid pattern of piling?

7) What is the main cause of cracking of walls and floors?

8) What is a pier made of?

(B) Foundation Settlements

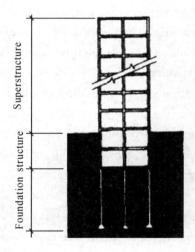
Superstructure, substructure and foundation

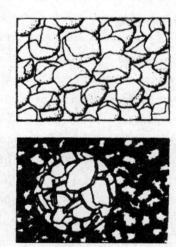
Silt particles(top) Clay particles(bottom)

It is convenient to imagine that a building consists of three main parts: the superstructure, which is the aboveground part of the building; the substructure, which is the habitable below-ground part and the foundations, which are the components of the building that transfer its loads into the soil.

All foundations settle to some extent as the soil around and beneath them adjust it to these loads. Foundations on bedrock settle a negligible amount. Foundations on certain types of clay, such as in Mexico City, may settle to an alarming degree, allowing buildings to subside by amounts that are measured in feet or meters. Foundation settlement in most buildings is measured in millimeters or fractions of inches. If settlement occurs at roughly the same rate from one side of the building to the other (uniform settlement), no harm is likely to be one to the building, but if large amounts of differential settlement occur, in which the various columns and load bearing walls of the building settle by substantially different amounts, the frame of the building may become distorted.

A primary objective in foundation design is to minimize differential settlement by loading the soil in such a way that equal settlement occurs under the various parts of the building. This is not difficult when all parts of the building rest on the same kind of soil, but can become a problem when a building occupies a piece of ground that is underlain by two or more areas of different types of soil with very different load bearing capacities. The following passage summarizes main types of soil for engineering purposes:

Rock is a continuous mass of solid mineral material, such as granite or limestone that can only be removed by drilling and blasting. Rock is not completely monolithic, but is crossed by a system of joints that divide it into irregular blocks. Despite these joints, rock is generally the strongest and most stable material on which a building can be founded.

Soil is a general term referring to earth material that is particulate. If an individual particle of soil is too large to lift by hand, or requires two hands to lift, it is known as a *boulder*. If it takes the whole hand to lift a particle, it is a *cobble*. If a particle can be lifted easily with thumb and fore-finger, the soil is gravel. In the Unified Soil Classification System, gravels are classified visually as having more than half their particles larger than 0.25 inch (6.5mm) in diameter. If the individual particles can be seen but are too small to be picked up individually, the soil is sand. Sand particles range in size from about 0.25 to 0.002 inch (6.5 to 0.06mm). Sand and gravel are considered to be coarse-grained soils.

Silt particles are approximately equi-dimensional, and range in size from 0.002 to 0.00008 inch (0.06 to 0.002mm). Clay particles are plate shaped rather equi-dimensional and smaller than silt particles, less than 0.00008 inch (0.002mm).

Clay particles, because of their smaller size and flatter shape, have a surface-area-to-volume ratio hundreds or thousands of times greater than that of silt. As particle size decreases, the size of the pores between the particles also decreases, and soil behavior increasingly depends on surface forces. The volume of the pores and the amount of water in the pores greatly influences the properties of a clay soil.

Peat, topsoil and other organic soils are not suitable for the support of building foundations. Because of their high content of organic matter, they are spongy and compress easily, and their properties can change over time due to changing water content or biological activity in the soil.

New Words and Expressions:

superstructure	n.	上部结构,上部建筑物
substructure	n.	下部结构,支撑结构,根基
foundation	n.	地基,房基
component	n.	组成部分,成分；零部件
adjust	v.	整理,整顿,使……适应
negligible	a.	不重要的,很微小的
fraction	n.	小部分
substantially	ad.	可观的,大量的
frame	n.	框架
minimize	v.	使降低到最低程度
capacity	n.	容纳某事物的力量
mineral	n.	矿物
granite	n.	花岗岩
limestone	n.	石灰岩
boulder	n.	（经风雨或水侵蚀而形成的）巨石
gravel	n.	砾石,沙子
coarse-grained	a.	粗糙颗粒的
approximately	ad.	大约的,大概的
ratio	n.	比率
peat	n.	泥煤,泥炭
topsoil	n.	表土,表土层
spongy	a.	海绵似的

Post-Reading Exercises:

1. Multiple Choices:

1) Foundations on bedrock settle _____.
 A. very little   B. negative   C. some   D. a few

2) A major _____ in foundation design is to minimize differential settlement by loading the soil in such a way that equal settlement occurs under the various parts of the building.
 A. aim   B. reason   C. object   D. challenge

3) If settlement occurs at _____ the same rate from one side of the building to the other, no harm is likely to be done to the building.

A. broadly B. nearly C. hardly D. sharply

4) This is not difficult when all parts of the building _____ on the same kind of soil.

A. hang B. are erected C. are used D. use

5) Rock is a completely _____ material, such as granite or limestone.

A. natural B. solid C. irregular D. durable

6) Silt particles are smaller than sand particles and range in size from _____.

A. 6.5 to 0.06mm B. 0.002mm
C. 0.2 to 0.06mm D. 0.06 to 0.002mm

7) The particle size and the size of the pores between the particles are _____ with each other.

A. closely connected B. inseparable associated
C. initially linked D. bound up

8) The properties of peat, topsoil and other organic soils can change over time _____ changing water content or biological activity in the soil.

A. because of B. because C. by D. therefore

2. Decide whether the following statements are true(T) or false(F) according to the passage given above:

1) The superstructure of a building is called foundation. (　　)

2) Foundations on certain types of clay show settlements of several millimeters. (　　)

3) Uniform settlements are settlements that occur at different parts of the foundation at the same time. (　　)

4) After drilling or blasting the inner structure of monolithic rock, it does not show irregular blocks. (　　)

5) Soil is a special kind of clay. (　　)

6) Soil is divided into several categories such as gravel, silt, sand or clay, according to the size of the particles. (　　)

7) Clay particles are smaller and flatter in shape than silt particles, and they have a surface-area-to-volume ratio ten times greater than that of silt. (　　)

8) Due to changes in the water content of the building's foundation, it is dangerous to use peat, topsoil and other organic soils as support of building foundations. (　　)

UNIT 6 WALLS

Part 1 Warm-up Activities

1. Read this passage and label the drawing below:

The external walls are made up of brick cladding, wall planks, windows, doors, heads and sills, stanchion casings and inner lining panels. While the steel frame is being erected, the wall planks and floor units are fixed. At the same time, the stanchions are enclosed in casings, which serve the function of resisting fire. The precast concrete floor units are capable of carrying a load of up to 5kn/sqm. The wall planks are designed to be weatherproof and to support the outer cladding. The aluminum heads, sills and windows are then fixed from inside of the building. After this, the 900mm and 1,800mm wide external doors are installed. These doors are either aluminum framed and pre-glazed or hardwood framed 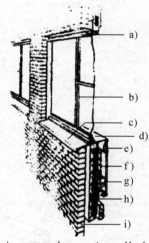 and the glazing is done on site. Finally, the internal sills and lining panels are installed. These form a cavity for the heating and electrical services. A grill underneath the sill, together with an air intake at skirting level, enables air to circulate up past the finned heating element. The lining panels are capable of being removed to give access to the services.

Now label this cross-section of the external wall of a building.

2. Now complete the following sentences to match the idea in brackets:

Example: The external walls are made up of brick cladding, wall planks, windows, doors, heads and sills, stanchion casings and inner lining panels.

a) The external walls... (structure).
b) The wall planks and floor units... (time).
c) The stanchion casings... (function).
d) The precast concrete floor units... (ability).
e) The wall planks... (function).

f) The external doors... (measurement).
g) The glazing of the hardwood framed doors... (location).
h) The internal sills and lining panels... (function).
i) The grill... (location).
j) The grill and air intake... (function).
k) The lining panels... (ability).

3. Answer the following questions:

a) Why do you think the aluminum heads, sills and windows are designed to be fixed from inside of the building?
b) What is the function of the fins on the heating element?
c) Are the aluminum-framed windows glazed on site?
d) What are the advantages of using aluminum instead of steel to make the windows?
e) What other types of cladding could be used instead of brick?
f) What could this type of building be used for?

Part 2   Controlled Practices

1. Thermal energy calculation:

Calculate the quantity of heat in Joules(J) that is flowing in 1hour(1h) through a brick wall of 0.2 meter(m) and, in comparison, through a timber wall of 0.1 meter(m).

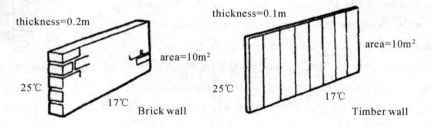

a) The ratio between the rates of flow of heat through the two walls is _____.
b) The brick wall has a _____ thermal conductivity than the timber wall.
c) The rate of flow of heat through a material is _____ proportional to its coefficient of thermal conductivity.
d) If we _____ the two materials, we can see that _____ is a relatively poor insulating material compared with _____.
e) We can tell that the material with a coefficient of thermal conductivity needs a relatively _____ wall to achieve the same degree of insulation.

2. Read the paragraph about the stability of concrete block walls and complete the following sentences to match the ideas in brackets:

Block walls should be designed so that they have stability against overturning. Walls may be divided into a series of panels and stability provided by connecting the edges of the panels to supports, which are capable of transmitting the lateral forces to the structure. The length or height of the panel in relation to the thickness of the wall has to be limited. The limits for three different design situations are described below and may not be exceeded.

    a) Block walls... (ability)
    b) The height of the panel... (proportion)
    c) The edges of the panels... (structure)
    d) The panel supports... (ability)
    e) If a wall is too thin... (cause and effect)

3. Learn about different design situations:

Design situation 1:

Walls with adequate lateral restraint at both ends but not at the top:

(i) The panel may be of any height provided the length does not exceed 40 times the thickness; or

(ii) The panel may be of any length provided the height does not exceed 15 times the thickness; or

(iii) Where the length of the panel is over 40 times and less than 59 times the thickness, the height plus twice the length may not exceed 133 times the thickness.

Design situation 2:

Walls with adequate lateral restraint at both the ends and the top:

(i) The panel may be of any height provided the length does not exceed 40 times the thickness; or

(ii) The panel may be of any length provided the height does not exceed 30 times the thickness; or

(iii) Where the length of the panel is over 40 times and less than 110 times the thickness, the length plus 3 times the height should not exceed 200 times the thickness.

Design situation 3:

Walls with adequate lateral restraint at the top but not at the ends:

The panel may be of any length provided the height does not exceed 30 times the thickness.

4. Identify the 3 design situations in the drawings below. Work out if any of the walls are unstable, because their design limits have been exceeded. Make sentences following the example:

Example: The 75mm solid block wall is stable(unstable), because its length does not exceed(exceeds) 40 times its thickness.

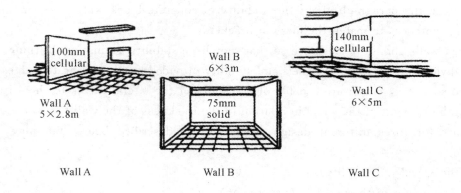

Wall A　　　　　　　Wall B　　　　　　　Wall C

Part 3　Further Development

1. Read and make an oral translation into Chinese:

Building a wall of natural stone is a hard work, and stone can be expensive. But the results are enduring and strikingly beautiful. Dry-laid stone walls rely on gravity rather than mortar, to hold them together. As a result, they are more difficult to build than mortared walls because you spend more time getting stones to fit well and rest solidly on each other. Don't try to build a dry stone wall more than 3 feet high; it is difficult to keep it stable.

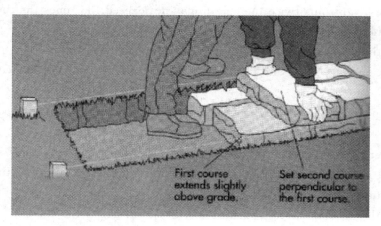

(1) Dig a trench and begin laying the stones.

Drive stakes and stretch mason's lines to establish the two outside edges of the bottom of the wall. Remove sod, any roots larger than 1/2 inch in diameter, and 2 or 3 inches of soil to provide a smooth and level base. Dig the trench deep enough so the top of the first course is slightly above grade.

Begin placing stones by setting them securely into the soil. Lay the stones in each course perpendicular to those in the course below them. This helps tie the courses together and strengthen the wall.

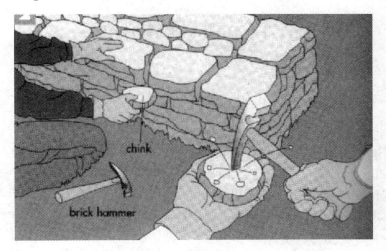

(2) Fill in voids with small stones.

Fill voids in the center of the wall with small stones and plug vertical gaps between stones by tapping chinks (small pieces of stone) into place. Cut these chinks and any other small stones you need with a brick hammer. Wear eye and hand protection when breaking stones. Always avoid placing stones of the same size directly on top of each other.

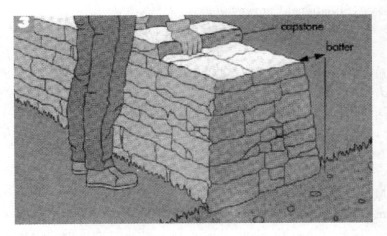

(3) Batter, top off with capstone.

Batter the wall as you work—that is, build the wall so it becomes slightly narrower as it rises. This makes the wall more stable. Notice in this wall how the stones form bonds both across the width and along the length of the wall. Select large, flat stones for the lop course. Some masons spread mortar over the stones in the next-to-top course, and then set the capstone into the mortar. This helps seal the top of the wall from moisture, which otherwise may freeze and weaken the wall.

Part 4　Business Activities

Possession of the Site

1. 文化与指南(Culture and Directions)

为了能使承包商顺利地开工(commence the work),在国际工程承包(international project contracting)合同中规定,若业主不能按规定提供施工现场,承包商有权获得相应的延期(extension of time)以及有关费用的补偿(cost compensation)。

由于能否及时进驻施工现场直接关系到承包商是否能顺利地实施工程,承包商在施工准备阶段(construction preparation stage)最重要的任务之一就是敦促业主将现场按时移交(hand over the site on time)。一旦业主未能按时向承包商移交现场,承包商应该在合同规定的时间内,向业主发出有关工期及费用的索赔通知(notification of claim),以弥补自己的损失。

同时,合理的现场布置有助于承包商高效率地进行工作(work efficiently)。生活营地(man camp)、材料储存场(storage area)、混凝土搅拌厂(batching plant)以及施工设备停放场(parking lot)的安排要以便利施工为标准。但现场的布置一般要征得业主的同意。

2. 情景会话(Situational Conversations)

(1) The Owner Hands Over the Site Late

To start the execution of works, the contractor needs to take possession of the site for the owner and plan the layout of the whole site for approval by the owner. The two parties are discussing the matters in handing over and arranging the site.

(A: The owner's representative, A_2: The owner's consultant; B: The contractor's project manager; B_2: Project manager'assistant from the contractor)

B: Mr. Madison, we sent you a letter last week, asking you to hand over the whole site to us, including the access road. But we haven't received a reply yet. The contract is quite specific about this in its special conditions: possession of the site shall be given to the contractor on the date named in the Appendix 1, which is April 20, 1992. Today is April 26. It's already one week late.

A: I'm sorry about this, Mr. Lin. Unfortunately we have met with some difficulty in requisitioning the land on the left bank of the river for the site areas. You know, some of the land is privately owned and the owners won't agree to sell the land. Nor do they want to grant us permission to use it because they are afraid that the project will disturb their peaceful life.

A_2: Some people here are obsessed with their traditional life; they don't want to have a change of their life! You'll never understand them.

B_2: It's so puzzling! They should know that they will get benefit from this project. At least they could get the lighting power easily, at a much lower price.

A: We've already promised to provide them with electric power for lighting free of charge. In return, they let us use their land for nothing.

B: When do you think you can solve the problem and make the whole site and the access road available to us? This is what we are concerned about.

A_2: At present, the access road has been built up to the dam site and we are beginning to build the road to the power-house from the junction. It happens to be the rainy season and the rains have slowed down our progress, but we are making a great effort to finish it soon.

A: So we can hand over to you the completed access road and the area on the right bank so that you can begin the preparation work there, you can have them from tomorrow. I'll give you a letter of confirmation right after this meeting. Once we solve our problem with the landowners, which I believe we can soon, we will hand over the remaining part of the site area.

B: Mr. Madison, let me make it plain to you. The delay in handing over the site area has adversely affected our construction plan. We have to request you to extend the completion time of this Project accordingly. Meanwhile, we reserve the right to be reimbursed for any costs incurred because of the delay. Fifteen engineers and technicians are already here waiting eagerly to start their jobs. A lot of preparation work needs to be done by us, especially the traffic road within the site.

A: I understand your position, Mr. Lin. As for the compensation, we'll try to settle this matter according to the contract.

B: To make this project a success, we need cooperation from each other.

A: I couldn't agree more.

(2) Laying out the Site

(To construct the works smoothly and efficiently, the site needs to be arranged properly. Today, the contractor's project manager and his assistant are talking to the owner's representative about the plan to lay out the whole site.)

(A: The owner's representative, A_2: The owner's consultant; B: The contractor's project manager; B_2: project manager'assistant from the contractor)

A: The other day, you told me that you were going to build up two camps on the site. Could you elaborate on your plan, Mr. Lin?

B: According to our construction plan, our first camp, which we call "Camp A", will be located at the junction of the two access roads leading to the dam and the power house. It will mainly be used for storage of construction materials, spare parts, etc. A parking lot and a repair shop will be set up within this area. The power bender will also be located here.

A: What will be the size of this camp?

B: It will cover an area of about 4,000 square meters.

A: How many is that in acres? I am not quite familiar with the Metric System.

B: Sorry, we are not used to the British System.

B_2: Just a minute. Let me work it out with my small calculator. Let's see, it's 4785 square yards, about one acre.

A: That's acceptable. This piece of land belongs to the government. We'll get it approved soon. What about the other camp?

B: The other camp, which we call "Camp B", is mainly for the accommodation of the Chinese professionals and the offices. It's also where we have our weekly and monthly review meetings. What do you call this kind of camp?

A_2: Man camp, to be specific, but it belongs to construction camp in general; or you could say it's part of the whole construction camp.

A: There's no clear distinction between these terms, even among professionals. Where will it be located?

B: On the right-hand side of the access road, about half a mile from Camp A. It will occupy approximately the same area.

A: Why do you have to choose this area, Mr. Lin? It is a part of this country's nature preserve. Though our government is in full support of this project, it will not sanction it at the price of damaging its beautiful environment. I don't think we could get the approval from the appropriate department.

B: You see, this area is ideal for Camp B. It's comparatively flat and this makes it easy for us to build the houses for bedrooms and offices. The key point is, it's close to the job site and the source of a potable water supply. Once we are given that piece of land, we'll be strictly confined to it and do everything we can to keep it in good order. Anyway, could you have a try with the government, please?

A: OK, if you insist. Will you submit to us all the layouts of the temporary works together with a letter of application? We'll try to get them approved soon, if possible.

B: The sooner, the better.

New Words and Expressions:

take possession of	占有,占用,进驻
access road	进场道路
be specific about	对……有明确规定
appendix n.	附录,附属物
requisition n.	征用

grant permission	给予许可
be obsessed with	对……着(入)迷
junction *n.*	(道路,河流等的)交叉处
letter of confirmation	确认函
remaining part	剩余时间
extend the completion time	延长竣工时间
reserve *v.*	保留
reimburse *v.*	补偿
incur *v.*	导致,引起
spare parts	备件
power bender	动力弯筋机
metric system	公制
British system	英制
acre	英亩($=4047m^2$)
man camp	生活营地
nature preserve	自然保护区
appropriate department	主管部门
be confined to	限制……的范围内
crushing system	人工沙石系统
commit oneself to	对……作出承诺,保证

Notes to the Text:

1. layout:布置图,计划。如:site layout(现场布置图),lay out 是动词词组,意思是"对……进行布置或定位",如:lay out the site(对现场进行布置)。

2. special conditions:特殊条件,专用条件。它是合同文件的一个重要组成部分,常被称为"第二部分"(Part II),一般条件或通用条件英语为 general conditions,常被称为"第一部分(Part I)"。

3. free of charge:免费。有时 free 单独使用时也有"免费的","免税的"意思,如:a free drink(一杯免费饮料),free discharge(免费卸货),free goods(免税进口的货物)。

4. for nothing:在这里也是"免费"的意思,相当于 free of charge。

5. make...available:意思是"提供……使其处于可使用的状态",相当于 provide and able to be used。例如:Water and power will be made available on the site by July 1.（工地在 7 月 1 号前通水通电。）

6. slow down:降低速度,慢下来。它既可以是及物的,也可以是不及物的。
例如:The car slowed down.（汽车降低了速度。）
Failure of the excavator to work efficiently greatly slowed down the progress of the excavation.（挖掘机工作效率不高大大地降低了开挖的速度。）

7. make...plain:将……讲清楚。这个词组中的 plain 意思是"明白的","清楚易懂的",等于 clear。例如:It is plain to everybody that the delay was caused by the flood event.

（大家都清楚延误是由洪水事件引起的。）

8. elaborate on：对……作进一步详尽的说明。该词组相当于 add more detail to。例如：Just tell us the facts, please. Don't elaborate on them.（请只告诉我们事实，不要多作解释。）

9. storage：储藏，保管。如：put the goods in storage（将货物入库），storage yard（储料场）。

10. work out：计算出，解决，制定。如：work out all the expenses（计算出所有开支）；work out a problem（解决一个难题）；work out a plan（制订一项计划）。

11. potable water：饮用水，也可以说 drinking water，"工业用水"的英语是 industrial water。

12. temporary works：临时设施，或者说 temporary facilities。

13. batching plant：混凝土搅拌厂，也可以说 batch plant。

Post-Reading Exercises：

1. "It happens to be the rainy season and the rains have slowed down our progress."（现在碰巧是雨季，下雨降低了我们的进展速度。）

"...happens to do (be) sth." "正巧……"

Now, you try doing the following：

1) 在开挖时，碰巧发现了一些古物。

(During the excavation, some antiques _____.)

2) A：Have you seen David recently?

B：昨天我在超级市场购物时正好见到他。

(I _____, when I was doing shopping in the supermarket yesterday.)

3) A：Was anyone injured in the cave-in accident?

B：很幸运，塌方时碰巧所有人员都在隧洞外面，所以没有造成人员伤亡。

(Fortunately, all the people _____ the tunnel when the cave-in occurred. So no one got injured.)

2. "...As for the compensation, we'll try to settle the matter according to the contract."（至于补偿，我们将按合同努力解决。）The phrase "according to" is used quite often during a negotiation, meaning "根据，按照"。

Now you try doing the following：

1) A：Why have you canceled the plan of tomorrow's concrete placement?

B：因为根据天气预报，明天将有大雨。

(Because _____, there will be a heavy rain tomorrow.)

2) 根据合同第十五条的规定，你方应于上月底之前向我方颁发该区段移交证书。

(You should have issued to us the taking-over certificate for the section by the end of last month _____ of the contract.)

3) A：Why do you reject the cement?

B：按照技术规范，所有运到现场的水泥都应为标准重量的密封袋装水泥。散装水泥不予验收。

(_____, all the cement shall be delivered to the site in sealed bags of standard weight. Bulk cement is not acceptable.)

3. "Meanwhile, we reserve the right to be reimbursed for any costs incurred by us because of the delay."(我方保留要求你方赔偿因耽搁而对我方造成费用的权利。)

"reserve the right to" is a very useful phrase used in administration of a contract, meaning"保留……权利"。

Now you try doing the following:
1)因为这件事属于不可抗力,我们保留对此提出索赔的权利。
(This event is force majeure, we _____ for it.)
2)这仅仅是协议草案,我们保留对其修改的权利。
(As this is just a draft agreement, we _____ modify it.)
3)我们保留随时撤回此项授权的权利。
(We _____ this delegation at any time.)

Part 5 Extensive Reading

(A) Casting A Concrete Wall

A reinforced concrete wall at ground level usually rests on a concrete strip footing. The footing is formed and poured much like a concrete slab on grade. Its cross-sectional dimensions and its reinforcing, if any, are determined by the structural engineer.

A key is sometimes formed in the top of the footing with strips of wood that are temporarily put into the wet concrete. The key is a groove that forms a mechanical connection to the wall. Vertically projecting dowels of steel reinforcing bars are usually installed in the footing before pouring. Later they will be over lapped with the vertical bars in the walls to form a strong structural connection. After pouring, the top of the footing is straight edged; no further finishing is required. The footing is left to dry for at least a day before the wall forms are erected.

The wall reinforcing is installed next, with the bars wired to one another at the intersections. The vertical bars are overlapped with the corresponding dowels projecting from the footing. At wall corners, L-shaped horizontal bars are installed to maintain full structural continuity between the two walls. If the wall is connected to a concrete floor or another wall at its top, rods will be left projecting from the top of the formwork to form a continuous connection.

Wall forms of lumber and plywood may be built for each section of the wall separately, but more often standard prefabricated formwork panels are used. The panels for one side of the form are coated with a form release compound, set on the footing, aligned carefully and braced. The form ties, which are small diameter steel rods shaped to hold the formwork together under the pressure of the wet concrete, are inserted through holes provided in the formwork panels. The ties will pass straight through the concrete wall from one side to another and remain in the wall. This may seem like an odd way to go about holding wall forms together, but the pressures of the wet, heavy concrete on the forms are so large that there is no other economical way of dealing with them. When the ties are in place and the reinforcing has been inspected, the formwork for the second side of the wall is erected. The forms are inspected to be sure that they are straight, plumb, correctly aligned, and adequately tied and braced.

Concrete is brought to the site and test cylinders are made to check for the proper pouring consistency. Concrete is then transported to the top of the wall by machines, such as a large crane-mounted bucket or by a concrete pump and hose. Workers standing on planks at the top of the forms, deposit the concrete in the forms, compacting it with a vibrator to eliminate air pockets. When the form has been filled and compacted up to the level that was marked inside the formwork, hand floats are used to smooth and level the top of the wall. The top of the form is then covered with a plastic sheet or canvas, and the wall is left to cure.

After a few days of curing, the bracing is taken down, the connectors are removed from the ends of the form ties, and the formwork is stripped from the wall. If required, major defects in the wall surface caused by defects in the formwork or inadequate filling of forms with concrete can be repaired at this time. The wall is now complete.

New Words and Expressions:

reinforced concrete	钢筋混凝土
dimensions n.	体积,大小,程度,范围
temporary a.	临时的,非永久的
groove n.	沟,槽,纹
overlap v.	部分重叠
erect v.	建造,设立,创立
install v.	安装
compound n.	复合物,化合物
diameter n.	直径
insert v.	插入,嵌入
consistency n.	强度,浓度
bucket n.	提桶,圆桶
brace n.	支撑,支柱
plumb n.	铅锤,测探锤
pump n.	泵
hose n.	软管

| deposit v. | 使沉积 |
| vibrator n. | 震动器 |

Post-Reading Exercises:

1. Translate the following phrases into Chinese:
1) a reinforced concrete wall
2) a concrete footing
3) steel reinforcing bars
4) to pour concrete
5) to fill the formwork with concrete
6) to compact the concrete
7) to cover the top of the form with a plastic sheet
8) the concrete is left curing
9) to strip the form work from the wall
10) to repair major defects in the wall surface

2. Decide whether the following statements are true(T) or false(F) according to the passage given above.

1) An architect calculates the cross-sectional dimensions and the required reinforcing of concrete footings.   (   )

2) A key made of steel is used as a groove that forms a mechanical connection between the footing and the wall.   (   )

3) Finally, the bottom of the footing is straight edged.   (   )

4) To form a continuous connection between walls and other structural members, steel rods will be left projecting from the top of the wall formwork.   (   )

5) It is necessary to check the formwork of the walls carefully before pouring concrete.   (   )

6) Concrete is transported to the top of the wall by hand.   (   )

7) It is necessary to protect the top of the newly erected walls while the concrete is drying.   (   )

8) After the concrete has been cured for a few weeks, the bracing is taken down, the connectors are removed, and the formwork is stripped from the wall. (   )

(B) Curtain Walls

In the evolution of structural systems, 2 basic types can be distinguished, deriving at various times according to particular circumstances. The systems are massive structures and skeleton structures.

In buildings of the first type, every part of the wall performs, without differentiation, the functions both of load bearing and of separation. In the second type on the other hand, a system of high strength units, which may be connected together in various ways, performs the special functions of a load bearing framework, while the other parts of the wall are devoted exclusively to the tasks of closing and separation. All non-load bearing walls, adopted at any time by whatever structural tradition, could in a certain sense be called curtain walls, In fact, in some respects they possess both the qualities of a wall(which is immovable, heavy and definitive) and those of a curtain(which is movable, light and temporary). In this broad sense, the Gothic cathedrals of Central and Northern Europe with their large windows between piers, the frame buildings of Japan with their panels of wood and paper may be considered early examples of curtain wall architecture.

However, the feature called by the modern name of curtain wall, signifies a particular kind of external, non-load bearing wall, composed of repeating modular elements, pre-fabricated or erected on site, which performs all the functions (and only these) of separation between indoors and outdoors, and in particular those of defense against external influences(atmospheric and otherwise), thermal and acoustical insulation, and regulation of view, light and air. This definition, however, is not always strictly adhered to, either because curtain walling is of recent invention, and has thus not been the subject of any deep critical examination, or because its typology is being constantly enriched, thanks to the continual efforts of designers and manufacturers to find better methods of production and application.

Curtain walls are the end product of a process of development that has involved a number of interrelated considerations connected with technical progress, social and cultural factors, and the emergence of the modern style in architecture.

The introduction and improvement of new structural techniques at the beginning of the 19th century, making use of steel, and in its second half, reinforced concrete, gave the impulse to a spreading use of framed structures. The increased use of this type of walling showed the importance of aiming at two characteristics in particular: Slenderness, to keep the maximum floor area available for use, and lightness, so that by reducing the load on the steel frame, the latter might be designed with correspondingly smaller members.

The small dimensions of the steel frames and the progress made by the glass industry permitted an increase "in window sizes, a development that was also stimulated by the demand for as much natural light as possible in industrial and commercial buildings. Between 1850 and the early years of the 20th century, the window gradually turned into the window wall, sometimes taking over the whole basic area defined by the facade. The use of large areas of glass had meanwhile become widespread in greenhouses and winter gardens, pedestrian galleries, railway station roofs and large exhibition pavilions.

The transformation of the window into the window wall and the employment of large glazed areas drew attention to a number of problems and evoked the first solutions to them. This included questions of insulation, eliminating condensation, developing the secondary glazing framework and considering the effects of expansion by carefully designing joints and casings. Large industrial buildings and tall office blocks, conceived as endless repetitions of identical cell units, led to the use of a uniform grid for the structural frames. Between the eve of the First World War and the beginning of the Second, the architects of the modern movement carried out a series of experiments, each of which may be considered as perfecting some particular aspect of curtain walling by the use of modern methods of industrial production.

At the same time, the theoretically principles were formulated by new architectural schools such as the "Bauhaus" in Germany and various projects by European architects such as Mies van der Roche and Walter Gropius. It was only after the Second World War, however, that the first built experiments emerged on a vast scale. It was in this period, that the curtain wall became that popular in the United States and Europe. Since then, architects have experimented with its application to various types of buildings, in close cooperation with the construction industry on the alert for new materials and methods.

Notes to the Text:

1. ... deriving at various times according to particular circumstances. The systems are massive structures and skeleton structures:……在特定环境下产生于不同时期。这些结构是整体结构和框架结构。
2. Gothic cathedrals:哥特式教堂。
3. the English term has universal currency:该英语名称世界通行。
4. composed of repeating modular elements:由同样的模数构件组成。
5. external agencies:外界各种侵蚀(指大气中的风、雨等)。
6. is not always strictly adhered to:并不自始至终被严格遵守。
7. gave the impulse to:促使(impulse 指"推动,冲动,冲力")。
8. evoked the first solutions to them:促使人们去率先解决这些问题(evoke 表示"唤起")。
9. on the alert for/against:密切注视,警惕。

New Words and Expressions:

curtain wall	幕墙,护墙
evolution n.	发展
distinguish v.	区别,辨别
civilization n.	文明,文化
exclusively ad.	专有的
possess v.	拥有,具有
plaster v.	粉刷,抹灰(泥)
partitioning v.	隔开
architecture n.	建筑
signify v.	预示
external a.	外面的

thermal	*a.*	热的
acoustical	*a.*	声的
typology	*n.*	象征意义的
reinforce	*v.*	增强
correspondingly	*ad.*	符合的,一致的
stimulate	*v.*	促进
pedestrian galleries		走廊
pavilion	*n.*	展出馆
condensation	*n.*	凝结物
aspect	*n.*	外观,方面
formulate	*v.*	配制
collaboration	*n.*	协作,合作

Post-Reading Exercises:

1. Multiple choices:

1) The _____ of curtain walls has had to do with the development of materials and methods.

　A. choice　　　B. evolution　　　C. increase　　　D. making

2) Reports that he is to be dismissed are gaining _____ among the staff members.

　A. currency　　B. popularity　　　C. credit　　　　D. belief

3) Students should strictly _____ to the rules and regulations of the school.

　A. abide　　　B. admit　　　　　C. advocate　　　D. adhere

4) We should never act on _____ but on reason.

　A. thought　　B. image　　　　　C. impulse　　　　D. will

5) The President's speech _____ great indignation from the public among developing countries.

　A. evoked　　 B. expressed　　　 C. rose　　　　　D. reacted

6) _____ is one of the advantages gained in building a skyscraper using curtain walls.

　A. Greatness　B. Heaviness　　　C. Width　　　　D. Slenderness

7) _____ the secondary glazing framework, curtain walls can ride behind or in front of the main structure.

　A. Thanks to　B. In spite of　　　C. Together with　D. Regardless of

8) The principles of governing the new city were _____ through collective efforts.

　A. formulated　B. manufactured　　C. installed　　　D. arrived

2. Answer the following questions:

1) What are the two basic types of structural systems?

2) According to the author, what kind of walls can fall into the category of curtain walls?

3) Why is the definition of the curtain wall in its modern sense not always strictly adhered to?

4) Why did framed structures become very popular and wide spread at the beginning of the 19th century and in the second half of the 19th century?

5) What are the chief advantages and problems of curtain walls?

6) When and why was the curtain wall developed and widespread in the U. S. and Europe?

UNIT 7 ROOFS

Part 1 Warm-up Activities

1. Look at these pictures, read aloud the labels and translate them into Chinese:

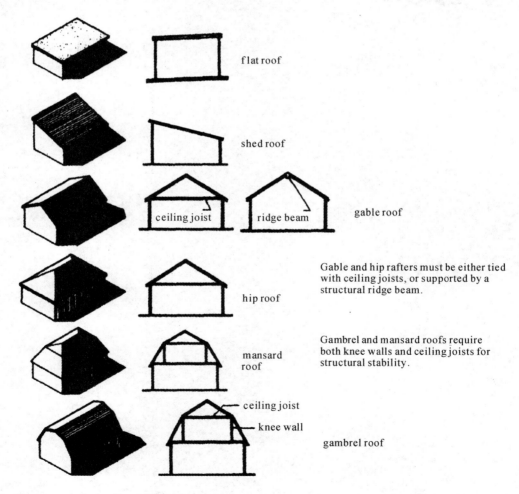

flat roof

shed roof

ceiling joist ridge beam gable roof

Gable and hip rafters must be either tied with ceiling joists, or supported by a structural ridge beam.

hip roof

Gambrel and mansard roofs require both knee walls and ceiling joists for structural stability.

mansard roof

ceiling joist
knee wall

gambrel roof

2. Learn about flat roofs:
Although termed flat roofs, they are often constructed with a slight fall to enable the

rainwater to run off. The main advantages of flat roofs are that they are comparatively simple to construct and generally less costly than pitched roofs. Now Answer the questions:

1) Is a flat roof really flat? Why is it constructed with a slight fall?

2) What are the advantages of flat roofs?

3. Learn about pitched roofs:

In the design of pitched roofs, one of the most important factors is the degree of the pitch or slope, which depends mainly on the material used to cover the roofs. The steeper the pitch, the more effective the roof is in quickly disposing of rainwater or snow. On the other hand, a steeper pitch entails a larger roof area and a higher cost. The roof space, commonly known as a loft, which is quite large in a roof with a steep pitch, is often used to provide additional rooms or a storage area.

Answer the questions:

1) Do you think that the steeper the pitch is, the better the roof is? Give your reasons.

2) Name the advantages and the disadvantages of pitched roofs.

Part 2   Controlled Practices

1. Read and do the exercises:

A simple flat roof consists of timber bearers covered with timber boarding and finished with several layers of tar felt. The first layer of tar felt is nailed to the boarding and the subsequent layers are bonded to each other in hot tar. The felt is then covered with stone chippings. The top layer of the felt at the vertical edges of the roof and against the wall is usually a heavy mineralized felt. As an alternative to tar felt, sheet zinc, aluminum or copper is often used in high class work as a finishing to a reinforced roof slab, but very seldom to timber flat roof construction.

Now please list the sequences to build a simple flat roof:

a) the first step is to _____.

b) the second step is to _____.

c) the third step is to _____.

d) the fourth step is to _____.

e) the fifth step is to _____.

Now please name the materials that can be used in making a flat roof as many as you can.

2. Look at this flat roof and translate the labels into Chinese:

① wall tile

② external skin

③ ventilation cavity

④ edging board

⑤ profiled coping

⑥ sheet metal flashing

⑦ two or three-layer tar felt

⑧ rough tongued and grooved boarding

⑨ roof bearer

⑩ binder

⑪ Layer of chippings

⑫ insulation quilts

⑬ lath

⑭ match boarding

⑮ internal skin

⑯ plastering

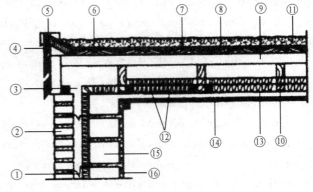

Flat Roofs

3. Look at this timber roof construction and translate it into Chinese:

The following is an example which illustrates an isometric view of a timber roof truss having a single post, common form of roof construction. The posts are spaced at intervals, varying according to the requirement of the design, and carry the ridge purlin, which is built into the gable walling at each end for additional support.

① gable wall

② verge

③ apex

④ post

⑤ chimney breast

⑥ strut

⑦ ridge purlin

⑧ chimney stack

⑨ chimney flue

⑩ roofing battens

⑪ rafter

⑫ wall plate, foot plate

⑬ ceiling joist

⑭ eaves

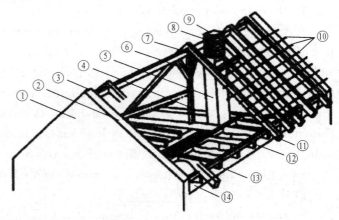

Isometric View of Roof Construction

4. Read this and list the assembly sequence to build a timber roof truss:

Struts are fixed to each side of the posts to provide further support to the ridge purlin. At the lower end of each roof slope, eaves purlins are provided over the ceiling joists. When they are actually fixed on top of the wall, they are known as wall plates. The rafters, spaced at approximately 80ctn centers, are nailed to the purlins.

  a) The first step is to _____ .
  b) The second step is to _____ .
  c) The third step is to _____ .
  d) The fourth step is to _____ .

Part 3   Further Development

1. Look at this gable roof and list the Chinese name of each component:

The following drawing illustrates a timber double post roof truss. This type of truss is constructed where a wider span or a less restricted roof space is required.

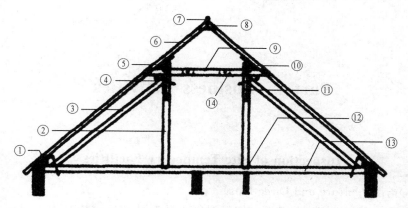

① wall plate
② post
③ brace, strut
④ bolted bridle joint
⑤ cleat
⑥ (common) rafter
⑦ ridge board
⑧ ridge gusset plate
⑨ tie beam
⑩ purlin
⑪ strut
⑫ subtenant
⑬ ceiling joist
⑭ packing piece

2. Read about a conventional built-up roof and sum it up with your own words:

For a conventional built-up roof, the membrane consists of four plies of felt bedded in asphalt with a gravel ballast. The base flashing is composed of two additional plies of felt that seal the edge of the membrane and reinforce it where it bends over the curb. The curb directs water toward interior drains or scuppers rather than allowing it to spill over the edge. The exposed vertical face of the metal roof edge is called a fascia.

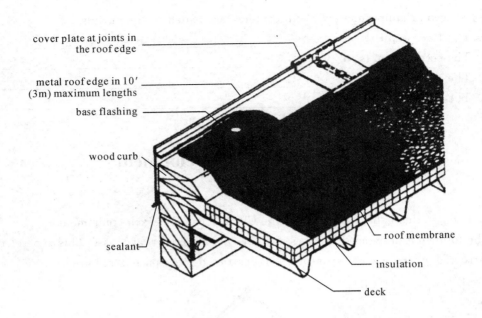

Part 4 Business Activities

Inspection of Site Temporary Facilities

1. 文化与指南(Culture and Directions)

建造承包商入驻工地后,由于许多钢筋混凝土预制件和钢结构必须在现场加工,施工现场需建造或设立许多临时设施(temporary facilities)。业主为了及时了解承包商的实力及工程建造情况,会定期或不定期地到现场检查是否具备这些临时设施以及这些临时设施的状况。这些临时设施包括:凝土搅拌站、钢筋场、钢筋混凝土预制场、木工房、机械维修间、电气车间、金属车间、水泥仓库,以及钢筋拉伸试验房、为混凝土测压的试验室、供电及给排水的临时设施。业主根据检查情况了解这些设施能否满足现场的需要并提出相应的整改要求。

2. 情景会话(Situational Conversation)

(1) Inspecting the Production Area

Weeks after the contractor signed the project contract and stationed the site, the first thing they need to do is to set up some temporary facilities, which is very necessary for the project to start well and develop safely. The owner of project himself, or his representative will come to inspect the site so as to make sure all the site temporary facilities are qualified.

(A: The owner's consultant engineer; B: representative of the contractor)

A: Now you have signed the contract for the project for a couple of weeks. We think it is necessary to have a check-up to your temporary site facilities.
B: Yes, we think so, too. Shall we go to the production area first? Here we are. This is the precast yard for the reinforcement concrete piles. These two gantry cranes are set for the piles handling.
A: How about the capacity of these cranes?
B: Each gantry crane in each production line can lift the weight of 10t. Here is the rebar yard. These bending machine and cutting machine can process rebar with diameter 40mm.
A: If I were you, I would use the automatic processing machine which is more efficient than these manual ones.
B: It will cost much more. As you know very well that our price was reduced by changing the advanced processing machines into manual one.
A: But at least you should have a shed for the yard. In such a hot area, if workers can work under a shed, it will certainly increase the efficiency.
B: It sounds reasonable. We will make a shed as a roof of the rebar yard. Here is the carpentry workshop. Please come in.
A: I am satisfied that every timber saw and plainer are equipped with a cover which is safe for the workers. But there is not enough equipment for fire extinguishing. You must increase at least 10 extinguishers in the workshop. And also you should put more slogans like "No Smoking" on the wall.
B: This is our machine repair workshop. It is divided into two parts: one for repairing heavy plants like excavators, bulldozers, crawler cranes, concrete pumps and so on, the other one for the vehicles including trucks and passenger cars. The gantry is five tone capacity. The garage is equipped much better than any one on the street.
A: Where is the electrical workshop? I hope you will not miss it.
B: Of course not. We will have at least 20 electricians working on this site for maintenance of the electrical instruments, cables and machines. The electrical facilities will be frequently set up and relocated. Here is the switch board. You see it is not a small one.
A: I see. It is equipped well. This one is no doubt a metal workshop. It seems quite big. How many workers work in this workshop?
B: Around 50. They are welders, fitters, machine operators and so on. From the drawings we know that there are a lot of steel embedments in concrete and many small steel structures such as ladders, rails, platforms and so on. All these things should be fabricated on site.
A: I think you should have a separate room for storing of the gas bottles and have specific workers to take care of them.

B: Yes, you are right. It has been scheduled to do that the next week. Here is the cement warehouse.

A: But this warehouse is not good for the storage of cement. First, the floor is wet because the floor level is not high enough for keeping dry. Second, the roof is not water proofed enough. You can see that there is a hole on the roof and the sunlight can come through in.

B: Thank you for pointing out these things to us. These should be improved before the cement is moved in.

A: Please let me know before your cement is stored in this warehouse.

B: OK. Now please see the concrete batching plants. The first batching plant with capacity 40m³/hour is ready for operation and the second is under erection.

A: According to the Method Statement you submitted, there will be three sets of batching plants for the project. When do you intend to erect the third one?

B: It will be ready in the end of the third month from now on. We plan to put the third batching plant mainly for stand-by and to operate within short time during the peak construction.

A: How big are these cement silos? Is the capacity enough for cement storage?

B: These two silos are only 200t capacity of each. We will build two more cement silos with 500t capacity for each.

A: That should be enough. Do you have an intention to install a chillier for the batching plant? Because it is very hot in summer and the temperature of the mixed concrete must be kept below the specification requirement, it will be very difficult to reach such condition without a chillier being equipped.

B: According to our study of the situation, we think that the concrete can be maintained within the required temperature by taking other measures. We will build a shed to cover the aggregate and sand without sunlight directly shining on.

A: I doubt if this method is effective or not. Your method can only be adopted before the summer comes. You should have an alternative way in case that this method can not maintain the concrete temperature complied with the specification.

B: We will keep you informed in different situation and prepare an alternative method for hot weather.

(2) A Visit to the Laboratory

A: Now I would like to have a visit to your laboratory. I hope that it will be a standard one since it is the very important facility you need for the project.

B: Here we are. This room is for concrete pressure test. There are hydraulic pressing machines for the sample crashing. Here are molders which are for both cubic and cylinder samples. The curing pools are built against the wall.

A: I appreciate you have prepared two sorts of molders in the laboratory. Do you have the instant concrete strength test instrument? It can be necessary for testing the hardened concrete.

B: It is not available at present. We will purchase two or three pieces soon.

A: What about the sieves? Can they deal with different materials?

B: Yes. These instruments are suitable for different sorts of material as long as the specification requires. These are the instruments for the soil tests. The dry density tests are operated very often for measuring earth compaction.

A: This room is no doubt for the steel bar tension test. It is good you have two machines equipped for the test. Do you have the equipment for the chemical analysis for steel in this laboratory?

B: Not yet at present. We will make these tests in the city laboratories because these tests are not often needed.

A: Now I would like to have a look at the site temporary systems for electricity, water supply and the drainage.

B: These systems are set up exactly in accordance with the layout showing during the tender works. This main transformer is connected from 100kV power line provided by the employer and supply 10kV power to the circuit surrounding our site.

A: Would you show me a transformer of 10kV/380V and a switch panel from this transformer?

B: Here we are. We have installed 10 nos. of this kind of transformers and around 20 nos. of the switch panels. Then the further step down switches will be connected to these panels. All these switches can be cut off automatically according to the timing limit when any short circuiting happens.

A: This sort of switch is very necessary for avoiding accident from electricity. Now we'd like to see the main pipeline of water supply.

B: The main pipe with a diameter of 150mm (millimeter) surrounding the site is connected from a pool on the hill. The branch pipes with valves are led to all places where water is needed. The water pool has 10,000m^3 capacity and is always kept full.

A: Now let's have a look at the drainage system. Do you separate the drains and sewers?

B: Yes, drainage and sewerage are separated. For sewer water, we built one septic tank for each toilet on site which will be cleaned regularly by a public sewer treatment company. The storm water is drained by cast iron pipes and open trench along side with the temporary road to the river. We think these systems can meet the demands of the site.

A: Generally speaking, I am satisfied with the site temporary facilities at the site which is established within such a short time. You should continue to improve these to make better use of for the works. That's all for today's inspection. Good bye!

B: Good bye!

New Words and Expressions:

temporary *a.*	临时的
site facilities	现场设施
production area	生产区
precast yard	预制场
reinforcement concrete piles	钢筋混凝土桩
gantry cranes	龙门吊
rebar yard	钢筋场
bending machine	弯筋机
timber saw	木工锯刨
extinguisher *n.*	灭火器
excavator *n.*	挖掘机
bulldozer *n.*	推土机
crawler *n.*	履带机
electrical facilities	电气设施
switch board	主开关箱
metal workshop	金属车间
welder *n.*	焊工
fitter *n.*	钳工
machine operator	机器操作工
fabricate *v.*	加工
gas bottle	氧气瓶
warehouse *n.*	仓库
batching plant	混凝土搅拌站
cement silo	水泥筒仓
chillier *n.*	制冷机
bar tension test	钢筋拉伸试验
valve *v./n.*	装阀门
drainage system	排水系统

Notes to the Text:

1. We think it is necessary to have a check-up to your temporary site facilities. 我们认为有必要检查一下你们的现场临时设施。

2. This is the precast yard for the reinforcement concrete piles. These two gantry cranes are set for the piles handling. 这是钢筋混凝土桩预制场。这两台龙门吊是供吊运桩的。

3. Here is the rebar yard. These bending machine and cutting machine can process rebar with diameter 40mm. 这是钢筋场。这些弯筋机和切断机可以加工 40 毫米直径的钢筋。

4. From the drawings we know that there are a lot of steel embedments in concrete and many small steel structures such as ladders, rails, platforms, and so on. 从图纸上我们知道混凝土中有许多钢埋件和许多小型的钢结构,如钢梯、栏杆、平台等等。

5. All these things should be fabricated on site. 所有这些东西必须在现场加工。

6. The first batching plant with capacity $40m^3$/hour is ready for operation and the second is under erection. 第一座 40 立方米/小时的混凝土搅拌厂准备投入使用,第二座正在安装。

7. We plan to put the third batching plant mainly for stand-by and to operate within short time during the peak construction. 我们计划将第三座搅拌厂作为备用,供高峰施工期间短期使用。

8. These two silos are only 200t capacity of each. We will build two more cement silos with 500t capacity for each. 这两个筒仓每个只有 200 吨的容量。我们将建两个以上容量为 500 吨的水泥筒仓。

9. It is very hot in summer and the temperature of the mixed concrete must be kept below the specification requirement, it will be very difficult to reach such condition without a chiller equipped. 夏天天气炎热,搅拌混凝土的温度必须保持不高于规范要求的温度,若没有一个冷却设备将很难达到这样的条件。

10. We will build a shed to cover the aggregate and sand without sunlight directly shining on. 我们要修建一个遮阳棚来避免骨料和砂直接被日晒。

11. You should have an alternative way in case that this method can not maintain the concrete temperature complied with the specification. 你们必须有一个替代方案,以防止这种方法不能保持规范要求的混凝土温度。

12. There are hydraulic pressing machines for the sample crashing. 这些是试块压力机。

13. Here are molders which are for both cubic and cylinder samples. The curing pools are built against the wall. 这是立方体和圆柱体试样模具。养护坑靠墙而建。

14. Do you have the instant concrete strength test instrument? It can be necessary for testing the hardened concrete. 你们有混凝土瞬时强度测试仪器吗?这是测试已硬化混凝土所必需的。

15. The main pipe with a diameter of 150mm (millimeter) surrounding the site is connected from a pool on the hill. The branch pipes with valves are led to all places where water is needed. The water pool has $10,000m^3$ capacity and is always kept full. 直径为 150 毫米的主管道与工地旁的一座山上的高位水池相连接,这个水池有 10000 立方米的容积,并且总是保持着盈满的状态。支管为闭合管网并配有阀门,能给所有需要用水的地方送水。

Post-Reading Exercises:

Translate the following sentences into Chinese:

1) I am satisfied that every timber saw and plainer are equipped with a cover which is safe for the workers. But there is not enough equipment for fire extinguishing. You must increase at least 10 extinguishers in the workshop. And also you should put more slogans like "No Smoking" on the wall.

2) This is our machine repair workshop. It is divided into two parts: one for repairing heavy plants like excavators, bulldozers, crawler cranes, concrete pumps and so on, the other one for the vehicles including trucks and passenger cars. The gantry is five tone capacity. The garage is equipped much better than any one on the street.

3) But this warehouse is not good for the storage of cement. First, the floor is wet because the floor level is not high enough for keeping dry. Second, the roof is not water proofed enough. You can see that there is a hole on the roof and the sunlight can come through in.

Part 5　Extensive Reading

(A) Roofs of A Building

A building's roof is its first line of defense against the weather. The roof protects the interior of the building from rain, snow and sun. The roof helps to insulate the building from extremes of heat and cold and to control the accompanying problems with condensation of water vapor.

And like any front-line defender, it must take the majority of the attack from the element: A roof is more exposed to the intense solar radiation than any other part of a building. At midday, the sun broils a roof with radiated heat and ultraviolet light. On clear nights, a roof radiates heat to the blackness of space and becomes colder than the surrounding air. From noon to midnight of the same day, it is possible for the surface temperature of a roof to vary from near boiling to below freezing. In cold climate, snow and ice cover a roof after winter storms, and cycles of freezing and thawing challenge the materials of the roof. A roof is vital to the sheltering function of a building, yet it is singularly vulnerable to the destructive forces of nature.

Roofs can be covered with many different materials. There are 2 main groups according to the slope of the roof: those that work on steep roofs and those that work on low-slope roofs, which are roofs that are nearly flat. The distinction is important. Due to the steep slope, water drains off quickly, giving wind and gravity little opportunity to push or pull water through the roofing material. Therefore, steep roofs can be covered with roofing materials that are fabricated and applied in small, overlapping units-shingles of wood, slate or artificial composition; tiles of fired clay or concrete; or even tightly wrapped bundles of reeds, leaves or grasses.

There are several advantages of these materials: many of them are inexpensive. The small, individual units are easy to handle and install. Local repair of damage is easy. The effects of thermal expansion and contraction, and of movements in the structure that supports the roof, are minimized by the ability of the small roofing units to move, with respect to one another. Water vapor vents itself easily from the interior of the building through the loose joints in the roofing material. And a steep roof of well-chosen materials skillfully installed can be a delight to the eye.

Low-slope roofs have none of these advantages. Water drains relatively slowly from their surfaces, and small errors in design or construction can cause puddles of standing water. Slight structural movements can also tear the membrane that covers the roof and keeps the water out of the building. Water vapor pressure from within the building can blister and damage the membrane.

If low-slope roofs are constructed properly, they will have some advantages. A low-slope roof can cover a building of any horizontal dimension, whereas a steep roof becomes uneconomically tall when used on a very large building. A building with a low-slope roof has a much simpler geometry that is often much less expensive to construct. And low-slope roofs can serve as balconies, decks, patios and even parks.

New Words and Expressions:

interior	*n.*	内部，里面
insulate	*v.*	隔离
condensation	*n.*	浓缩
vapor	*n.*	潮气
ultraviolet	*a.*	紫外线的
vulnerable	*a.*	无防御的
destructive	*a.*	受到毁坏的
slope	*n./v.*	斜线；倾斜
shingle	*n.*	木瓦，屋顶板
slate	*n.*	石板
artificial	*a.*	人工的
clay	*n.*	黏土
concrete	*n.*	混凝土
thermal	*a.*	热的；保温的
contraction	*n.*	缩小
puddles	*n.*	水坑
membrane	*n.*	膜

blister *n.*	水泡,起泡	
horizontal *a.*	水平的	
uneconomically *ad.*	不经济地,浪费	
balcony *n.*	阳台	
deck *n.*	甲板	

Post-Reading Exercises:
Multiple Choices:

1) A roof is always the _____ part of the house.
   A. coldest   B. warmest   C. most exposed   D. best covered

2) During the day, the sun heats up the roof, but during the night, the surface temperature of the same roof may _____ below zero.
   A. drop   B. increase   C. rise   D. climb

3) A roof is _____ to the sheltering function of a building, yet it is singularly vulnerable to the destructive forces of nature.
   A. essential   B. helpful   C. useful   D. important

4) Steep roofs can be covered with roofing materials that are fabricated and _____ in small, overlapping units.
   A. shaped   B. decorated   C. arranged   D. set

5) The ability of the small roofing units to move with respect to one another _____ the effects of thermal expansion in the structure that supports the roof.
   A. increases   B. improves   C. decreases   D. leads to

6) And a steep roof of well-chosen materials skillfully installed can be _____ to the eye.
   A. inspiring   B. pleasant   C. dangerous   D. harsh

7) Structural movements can also tear the membrane that covers the roof and keeps the water out of the building.
   A. big   B. strong   C. little   D. less

8) A low-slope roof can cover every building no matter how _____.
   A. tall   B. heavy   C. wide   D. thick

(B) Roof Construction

The main purpose of a roof is to cover the upper parts of a building as a protection against the elements, such as wind, rain and snow. In dwelling houses, industrial and commercial structures, special attention is focused on fire-resistance measures of roofing.

Roofs may be flat or pitched. Covering materials for pitched roofs are selected according to their resistance to rain and snow, their durability in changes of temperature and in acid atmospheres, and for their artistic qualities in relation to architectural design. Nowadays due to economical reasons, the pitched roof has given way to the flat roof in modern office buildings. Flat roofs are usually covered with tar, metal or gravel. Nevertheless sufficient fall must be given, so that water can drain off quickly and adequate gutters have to be placed in the concrete of a flat roof to discharge the water in heavy storms.

Timber framework continues to be used for supporting roof coverings of lead, copper, zinc and tiles—both for flat and pitched roofs. Steel has largely replaced timber for the main structure of a roof and for many transverse members in large buildings and even for semi-permanent industrial buildings and stores, because of the rapidity with which materials can be prepared, assembled and erected. From an economically point of view, steel may also be the better choice in roof framing except for dwellings and small buildings, or for dye-works and bleak works, where it is impossible to use steel because of its rapid oxidation. Roofs are often made with considerable projection at the eaves to protect the upper portion of the external walls. To increase the esthetic value in design might be another reason.

Flat roofs are simple and widely used to cover large rectangular buildings. They consist of joints or horizontal beams, supporting a cover of boarding over which a membrane of weatherproof material is laid. Flat roofs in industrial construction use exceptionally heavy timber beams as a fire-control measure. Usually they receive additional support from X bracing, called bridging and between the joists, interior columns support such massive roofs.

Single-slope roofs are less often used than double-slope or gabled roofs. The principal components of the latter are pairs of rafters and light supporting beams. The beams meet in a line of inverted V's at a horizontal ridge beam and are connected below by the floor of the top storey, forming a truss. A large gabled roof structure requires additional strengthening, which may be achieved by vertical columns supporting the ridge beam at the junctures. Another method of support for very large roofs is the vertical support provided at an intermediate point of each rafter: Each pair of posts is tied together at the top by a horizontal straining beam.

In modern buildings, large timber trusses with strong connections can support gabled roofs with spans of 200 feet (60 meters) or more. For such spans, however, steel or reinforced concrete is commonly used. Roof coverings for certain very large modern buildings, such as sports arenas, have developed a variety of innovative designs, such as bi-grid (or tri-grid) space-framed steel structures suspended by cables from a high exterior wall. Modern roofing materials include aluminum and zinc, as well as such older a materials as copper and tiles.

New Words and Expressions:
dwelling house	住宅
fire-resistance n.	防火

incline n./v.	倾斜,斜坡
pitch n./v.	倾斜,倾状
durability n.	耐久性
acid a.	酸性的
artistic a.	艺术的
prevail v.	流行,盛行
tar n.	焦油沥青
asphalt n.	柏油沥青
drain n./v.	排水,阴沟
gutter n.	水槽
discharge n./v.	排出
copper n.	铜
zinc n.	锌
slate n.	板盖;铺石板
tile n.	盖瓦
pitched roof n.	坡顶
transverse a.	横断的
semi-permanent a.	半永久性的
dye-works n.	印染厂
bleach-works n.	漂白厂
oxidation n.	氧化
projection n.	投射,投影图
rectangular a.	矩形的,直角的
membrane n.	薄膜,隔膜
weather-proof a.	不受气候影响的
fire-control a.	防火的,耐火的
X-bracing n.	交叉支撑
single-slope a.	单坡的
lean-to a./n.	单坡的(房子);披屋
gable n.	山墙
gable roof n.	人字屋顶,三角形屋顶
rafter n./v.	椽;装椽子
invert v.	颠倒;错位
ridge n.	屋脊
intermediate a.	中间的
vertical a.	垂直的

Post-Reading Exercises:

1. Translate the following expressions:

1) a covering of a structure
2) protection against wind, rain and snow
3) in dwelling house
4) at an angle to suit the roof covering
5) durability in acid atmosphere
6) modern office buildings
7) timber framework
8) adequate gutters formed in the concrete
9) transverse members
10) dye-works and bleach-works
11) possess aesthetic value in design
12) supported by interior columns
13) in lean-to construction
14) vertical support at an intermediate point
15) a horizontal straining beam
16) 装配和吊装
17) 保护外墙上部
18) 坡顶屋面材料
19) 屋顶的檐口处
20) 薄膜
21) 大型三角架结构
22) 轻型支承梁塔
23) 特种油毛毡石棉板
24) 接点坚固的木桁架
25) 现代屋面材料

2. Put the following into English:

1) 屋顶的主要用途是封闭房屋的上部以防风、雨和雪。

2) 平顶的建造比坡顶的建造更为经济,为此,坡顶已让位于平顶。

3) 钢制构件能够高速制作、装配和吊装。

4) 支撑大型屋顶的方法是在每根椽子的中部支一根立柱。

5) 跨度为两百或三百英尺以上的三角屋顶一般都采用钢结构、钢筋混凝土结构或预应力混凝土结构。

UNIT 8 TOOLS AND METHODS

Part 1 Warm-up Activities

1. Look at these diagrams of the tools used by tradesmen working on a building site:

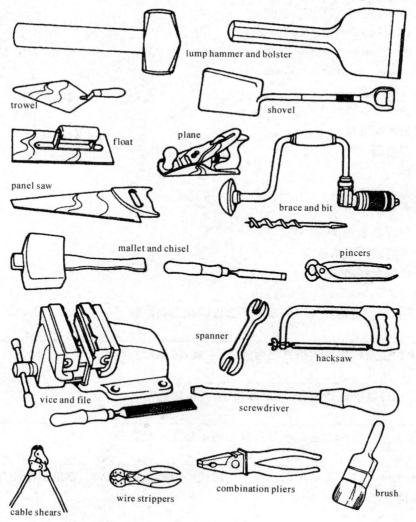

2. Now complete the following table with the correct tools or combination of tools for the jobs and make sentences as the example:

Example: A brace and bit is a tool for a carpenter to drill holes in wood.

Tradesman	Job	Tool
Carpenter	drilling holes in wood	
Bricklayer	mixing mortar	
Plasterer	smoothing the plaster on a wall	
Carpenter	cutting wood	
Plumber	cutting metal pipes	
Electrician	cutting electric cables	
Carpenter	making joints	
Plumber	smoothing metal surfaces	
Electrician	removing the outer sheathing of wire	
Carpenter	turning screws	
Decorator	painting surfaces	
Bricklayer	cutting bricks	
Plumber	tightening nuts	
Electrician	twisting strands of wire together	
Carpenter	smoothing wood surfaces	
Bricklayer	laying mortar on bricks	
Carpenter	removing nails	

3. Look at these drawings of instruments and make sentences:

Example: Verticality can/may be checked by using(or with) a plumb bob.

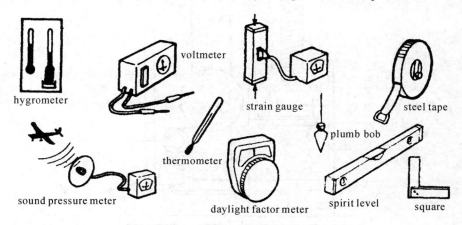

4. Now make sentences from this table:

A/An	lighting engineer structural engineer bricklayer acoustic engineer carpenter electrician	uses	a square a plumb-bob a hygrometer a strain gauge a voltmeter a spirit level a steel tape a thermometer a daylight factor meter a thermometer	to	check verticality. measure the illumination from the sky. measure the sound pressure. measure the relative humidity. check vertical and horizontal work. measure the temperature. measure the voltage of a circuit. check squareness. measure distances. measure the strain on a structure.

Part 2　Controlled Practices

1. Read about an experiment to investigate the effect of water content on the compressive strength of concrete:

Three different mixes of concrete were separately prepared. The materials in mix A were mixed dry in the proportion of 1∶1∶2. The cement used was normal Portland cement. The mix was divided and each half was separately mixed with water, one half with 30 percent more water added than the other. A 150mm cube was then made from each half of each batch and tested for compression strength at the end of twenty-eight days. Mixes B and C were dealt with in a similar manner, the excess water added to one-half of each mix being also 30 percent.

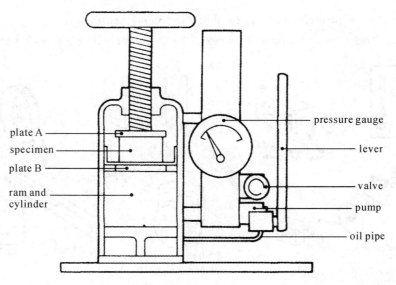

Compression testing machine

2. Calculation of reduction in strength for concrete mixes:

	Mix Proportions of cement : sand : aggregate	Normal mixes		Mixes with 30 percent excess water	
		water-cement ratio	Strength (N/mm²)	water-cement ratio	strength
A	1 : 1 : 2	0.43	34.9	0.56	23.4
B	1 : 2 : 4	0.62	20.3	0.81	11.6
C	1 : 3 : 6	0.85	10.7	1.10	5.9

From these results we can calculate the percent reduction in strength for the 3 mixes due to the excess water. Now draw a table and calculate this reduction in strength for mixes B and C(33 percent reduction in strength).

Now fill in the missing information:

We can conclude that the _____ of concrete is considerably _____ as a result of the additional _____. The reason for this is that water combines chemically with cement and an excess of water weakens this reaction on which the strength of the concrete depends.

3. Read these instructions for carrying out a compression test on a concrete cube using the apparatus shown in the diagram in exercise 1:

a) Cast the concrete mix in a steel mould of 150mm × 150mm × 150mm internal dimensions and store in a damp cabinet for 24 hours.

b) Remove the specimen from the mould and immerse in water until the cube is ready for testing.

c) Place the concrete specimen on the lower compression plate.

d) Lower the upper compression plate bin to top of the specimen.

e) Raise the lower lever to operate the hydraulic ram moving plate B upwards.

f) Continue this pumping action until the specimen is crushed.

g) Note the reading on the pressure gauge.

Now look at these results of compression tests done on concrete specimens made of three different types of cement and performed at different intervals of time:

Type of cement	Nominal mix	Compressive strength(N/mm²)				
		3 days	7 days	28 days	3 months	1 year
Ordinary Portland		9.65	17.2	26.8	33	45
Rapid-hardening	1 : 2 : 4	17.2	24.1	34.4	38	48.2
High alumina		48.2	49	55	No appreciable increase	

Write a report of the experiment with the following headings:

Purpose   Results
Description of apparatus   Conclusion
Procedure

Part 3　Further Development

Read the following sentences and learn them by heart.

1) There are many tools in the tool storage unit(tool chest, tool box). The tools must be well kept. 工具柜(工具盒、工具箱)里面有很多工具。必须妥善保管好工具。

2) Get me a hammer(hacksaw, file, scraper, chisel, socket wrench, hook spanner, adjustable wrench, pipe wrench, ratchet wrench, open end wrench, screw driver, hand vice, pliers and pocket knife). 给我拿一把手锤(钢锯、锉刀、刮刀、凿子、套筒扳手、钩扳手、活动扳手、管扳手、棘轮扳手、开口扳手、螺丝刀、手钳、扁嘴钳、小刀)。

3) Straightedge rule(square rule, slide gauge, inside and outside micrometer, steel tape, feeler, dial gauge, depth micrometer, wire gauge and radius gauge and thread pitch gauge) is a kind of common measuring tool. 直尺(角尺、游标卡尺、内径和外径千分尺、钢卷尺、塞尺、千分表、深度千分尺、线规、半径规、螺距规)是一种常用量具。

4) The precision of this fitter level(cross-test level) is 0.02mm/m. 这个钳工水平仪(框架式水平仪)的精度为0.02毫米/米。

5) We have got the instrument(pressure gauge, thermometer, tachometer, current meter and universal meter) ready for the experiment(test). 我们已经准备好做实验(试验)的仪器(压力表、温度计、转速计、电流表、万用表)。

6) That is an air(electric) powered grinder(portable grinder, angle grinder, straight grinder, drill, impact wrench, riveting hammer, hammer drill). 那是一个气(电)动砂轮机(手持砂轮机、角型砂轮、直型砂轮、钻机、冲击扳手、铆钉锤、锤钻机)。

7) Our electrical tools are double insulated and approved to international safety standards. 我们的电动工具都是双重绝缘的，符合国际安全标准。

8) Hydraulic pump is the power unit of the hydraulic puller(hydraulic press, hydraulic pipe bender, hydraulic jack). 油压泵是油压拉出器(油压机、油压弯管机、油压千斤顶)的动力装置。

9) A welder's kit contains electrode holder, welding torch, helmet shield, portable electrode heating box and temperature measuring pen. 一名焊工的成套工具包括焊钳、焊炬、面罩、手提式焊条加热箱和测温笔。

10) The diameter of this wire rope(hemp rope, sling) is three-fourth inches(3/4″). 这钢丝绳(麻绳、吊索)的直径为3/4英寸。

11) The lifting capacity of this chain hoist(hydraulic jack, screw jack) is 5 tons. 这个吊链(油压千斤顶、螺旋千斤顶)的起重能力为5吨。

12) The vise(parallel-jaw vice) is available to all of the bench work. 所有的钳工工作都可使用台钳(平口钳)。

13) Grease gun and oiler are the lubrication service tools for machinery. 油枪和注油器

都是机械润滑维护工具。

14) Torque wrenches offer the precision measurement needed to tighten fasteners. 力矩扳手可以提供紧固螺栓所需的精确力矩计量。

15) The measuring unit of torque wrench is pound-inch or kilogram-centimeter. 力矩扳手的计量单位为镑—寸或者千克—厘米。

16) Is the machine accompanied with some tools(spare parts, accessories)? 这台机器随机带有一些工具(备件、附件)吗？

17) Shall we use a special tool for this job? 我们干这活要使用专用工具吗？

18) Could you tell us how to use(operate, repair, maintain, clean, adjust) this new tool? 你能告诉我们如何使用(操作、修理、维护、清理、调整)这个新工具吗？

19) The tool gets out of order, and we must remedy its trouble. 这工具有毛病，我们必须排除它的故障。

20) The tool is out of repair, it needs an overhaul. 这个工具破损了，需要拆修。

Part 4　Business Activities

Procuring the Goods

1. 文化与指南(Culture and Directions)

做好物资采购(procurement of goods)工作是搞好国际工程承包的重要环节之一。这不仅因为材料与设备(materials and equipment)在工程合同中占有很大的比重，而且其采购与供应直接关系到工程的进度与质量。因此，一个有经验的国际承包商(experienced international contractor)往往十分重视工程承包中的物资采购工作。

在投标前，承包商一般要在工程所在国及邻国进行市场调查(market research)，寻找中标后工程所用物资的货源。中标后，应根据施工组织设计(construction management plan)，编制物资供应总计划，交物资部门(purchasing department)组织供应。物资部门接到计划后实施设备材料的采购工作。

采购的第一步是询价。询价的方式一般用电子邮件或传真向几家供货商发出询价函，然后"货比三家"，从中择优。询价时一定要将材料或设备的规格写清楚，以防误购，并因此耽误工程的施工。对于大量的、重要的以及定期的采购(substantial, important and regular purchase)最好派专人进行口头谈判(verbal negotiation)，以争取优惠的供货条件(special terms)。

租赁(leasing and hiring)也是施工设备(construction equipment)供应渠道之一。施工设备的租赁一般用于两种情况：一种是在工程前期，在所购设备还没有到达现场之前，为了满足紧急开工的需要而租赁；另一种情况是，由于有些设备虽然是施工所必要的，但使用时间短，利用率低，对于此类设备也常常采用租赁方式。

2. 情景会话(Situational Conversations)

(1) Making an Enquiry

To get the project under way, a lot of equipment and materials are required. All these have to be purchased or hired and delivered to the construction site. The purchasing clerks of the contractor make enquiries, negotiate supply contracts and, sometimes, hire equipment that is needed for a short period only. If the goods are imported from a third country or the home country of the contractor, they will have to go through the customs clearance formalities and, therefore, a customs by the contractor.

(A: manager in charge of purchasing the goods for the project; B: sales manager of the corporation about some construction materials)

B: Hello, Mr. Huang. How nice to see you again!

A: Nice to see you again, too, Mr. Chamberlain.

B: When did you come back to this country?

A: Only last week. How is the business?

B: Pretty good. I hear that COCC has won the contract to construct the hydro works. Is that true?

A: You're well informed. Yes, that's why I am back. When I came here to study the local market for the preparation of our bid last November, I told you that I would come again for a detailed business talk with you once we got the contract. Now, I am here again.

B: We really appreciate your visit, Mr. Huang. Our corporation has just expanded business operations, thanks to increasing demand and our good reputation. In addition to steel and lumber, we now also deal in such construction materials as cement, bitumen and PVC pipes. Here is our new catalogue.

A: Thank you.

B: Do you have any idea of what to order from us?

A: Let me see. Please quote us for Catalogue No. 12, that is, Grade 60, A615 hot rolled corrugated steel bars of size $3/8'' \times 40'$. Remember that if the price is acceptable to us, we'll place an order of not less than 50 tons for the first delivery.

B: I know you're a big potential customer for us. We will certainly give you a most favorable price even for the first deal. I believe we will be able to build a good business relationship with each other.

A: Can you provide the technical date?

B: No problem. The technical data are always available from the laboratory of the manufacturer.

A: To be frank with you, I've enquired of several local trading companies and I'm also waiting for their quotations, but I hope you will be chosen as our long-term supplier.

B: Thanks. I'll send you our quotation sheets in two days. Although we are not the biggest supplier in this line, well prove to you that we are the best.

(2) Hiring Equipment

(A: *the project manager*; B: *equipment manager of construction holdings*)

A: I have been told, Mr. Boles, that you are in charge of the construction equipment leasing and hiring business in this company. I am here to see if I can hire some equipment from you.

B: You've come to the right place, Mr. Yang. We are involved primarily in equipment leasing and hiring as well as mining operations. We own quite a lot of heavy construction equipment and can hire them to you at any time you like.

A: What are they?

B: We have bulldozer, excavators, loaders, motor graders, vibrating rollers and dump trucks.

A: I am thinking of hiring an excavator and a loader this time. What's the volume of your excavator's bucket?

B: The excavator is a 225 Caterpillar and has a bucket of about two and half cubic yards.

A: And the loader?

B: It's a 9508 Caterpillar. I think the volume of the bucket is about three cubic yards.

A: What are their hire rates respectively?

B: Where do you intend to use them?

A: They will be used to excavate the foundation pit of the powerhouse on the job site. Blasting will be done before the excavator is used. What the excavator dig is just small broken pieces of the rock. The working conditions are not so bad.

B: Well, Mr. Yang, we have two rates for you. The hourly rate is US $75 for the excavator and US $50 for the loader. The monthly rate is US $22,500 for the excavator and US $150,00 for the loader.

A: Does the rate include the fuel and the operator?

B: Yes, and maintenance as well.

A: How about the rate if we provide the fuel and the operator?

B: As a rule, we have our own operators and don't allow other operators to use our machines. You know, the machines are very expensive. I hope you understand. However, if you provide the fuel free of charge, we can reduce the rate by US $5 for each hour or US $1,500 for each month.

A: Who will be responsible for the transportation of the equipment from your plant depot to our job site?

B: If the hire term exceeds 300 working hours or one working month, we will transport the equipment by trailer at our own cost. Otherwise, we will charge US

$200 for the transportation of each machine.

A: How do you usually record the working hours?

B: Our operator fills in the time sheets which are countersigned by your site superintendent. We have set forms of time sheets. Every time the sheet will be in duplicate, one copy is kept by each side. Payment will be made according to the time sheets.

A: Talking about payments, how often do you want them to be made?

B: At the end of each working week.

A: We prefer to pay you monthly because we get payment from the project owner every month. We plan to use your equipment for two months this time. It's not a short period.

B: All right. Do you mean you will hire more equipment in the future?

A: I would say yes and, probably, on a long term basis if the current rates could be reduced by 20 percent. The quantity of our excavation work has increased due to the change of the construction design. We're now thinking of buying or hiring additional equipment. Judging by our calculation, it is more cost-effective to hire equipment if the hourly hire rate, say, an excavator, could be around US $60.

B: To show our friendship and goodwill to establish good relationships with you, I am willing to give a five percent discount of the rates I quoted you for the excavator and the loader. If you hire any equipment from us in the future for three months or more, we will reduce the rate by fifteen percent.

A: Thank you. If everything is all right, we want the machines to begin working by next Thursday.

B: No problem. As soon as we receive your notice and confirmation of our talk today, we will deliver the machines to you.

New Words and Expressions:

get the project under way	开始执行项目
purchasing clerk	采购人员
make enquiries	询价,询问
negotiate v.	谈判,协商
supply contract	供货合同
go through	通过,完成
formality n.	手续
customs agent	海关代理
sales manager	销售经理
local market	当地市场
catalogue n.	(产品)目录
place an order	订货

favorable price	优惠价格
construction design	施工计划
maintenance *n.*	保养
time sheet	计时单
countersign *v.*	会签,副署
set form	固定格式
in duplicate	一式两份
current rate	现有价格,时价
cost-effective *a.*	经济的,合算的

Notes to the Text:

1. home country:(承包商)本国。home 在这里的意思是"国内的",如:home office(公司总部),home-made(国产的),home equipment(国产设备)。

2. enquire of sb. about sth. :"向……询价"。

3. You're well informed! 你的消息可真灵通!

4. thanks to:由于。在这里相当于 owing to, because of。

5. PVC pipe:PVC 管子。PVC=polyvinyl chloride(聚氯乙烯)。

6. hot rolled corrugated steel bars:热轧竹节钢筋。

7. leasing and hiring:这是两个同意词组合,同为"租赁"之意。

8. bucket:铲斗,挖斗。这个词还常有"(水)桶"的意思。

9. a 225 Caterpillar:一台 225 型 Caterpillar 挖掘机。Caterpillar 后面省略了 excavator 一词,下文的 a 9508 Caterpillar 后面同样省略了 loader 一词。Caterpillar 为一家机械设备公司的名字。

10. plant depot:设备存放场。depot 意思是"库","基地",如:fuel depot(油库),repair depot(修理厂)。

Post-Reading Exercises:

1. "Please quote us for Catalogue No. 12..."(请给我们报产品目录第 12 号的价格。)
"So quote us the rate for bulk cement, please."(请给我们报散装水泥的单价。)
"quote sb. (the price/rate)for sth." "请给……报……的价格/单价。"
Now you try doing the following:

1) A: Our company deals in various construction materials.
B: Please _____ for all-in aggregate.(请给我们报毛骨料的价格。)

2) A: Please _____ for this kind of mosaic.(请给我们报这种马赛克的价格。)
B: What's the quantity you are going to order?

3) A: We can make windows of this kind according to your specification.
B: Please _____ the lump sum for one hundred set.(请给我们报 100 套的总价。)

2. "... we're in urgent need of 500 tons of each of Types 1 and 5 of Portland

Cement."(……我们急需一型和五型硅酸盐水泥各500吨。)

be in urgent need of sth. :"急需……"

Now you try do the following:

1) A: We're _____ 10 tons of galvanized steel sheet. I wonder if you could deliver it within one week.(我们急需10吨镀锌钢板,不知你方能否一周内交货。)

B: No problem.

2) A: We can supply the course sand at the price of 10 dollars per cubic yard.

B: It is true that we _____ it, but I'm afraid the price is too high for us to accept.(虽然我们急需,但恐怕价格太高,我们无法接受。)

3) A: This disease is contagious and the patient _____ medical treatment.(这种病属于传染性疾病,病人急需治疗。)

B: How long will he be hospitalized?

3. "…if you provide the fuel free of charge, we can reduce the rate by $5 for each hour."(……如果你们免费提供燃料,我们可以每小时降低5美元。)

"If you hire any equipment from us in the future for three months or more, we'll reduce the rate by 15 percent."(如果今后你们从我方租用任何设备的时间达到或超过三个月,我们将把租费降低15%。)

"reduce the price/rate by...":"把价格/费率降低……"

Now you try doing the following:

1) A: This is not the first time that we buy the asbestos-cement board(石棉水泥板) from you. I hope you can give us a favorable price.

B: All right. We'll _____ 3 percent for your order this time.
(好吧,对于你们这次的订购,我们降价3%。)

2) A: If you _____ 10 dollars, we have to make the purchase from other manufacturers.(如果你们不将此价格降低10元,我们只得从别处购买了。)

B: I'm afraid to say you're asking too much.

3) A: As a rule, we _____ 5 percent to regular customers. I hope you could come often.(我们对常客一般降价5%。希望您常来。)

B: Thank you. I'll keep it in mind.

Part 5 Extensive Reading

(A) Calculate the Amount of Daylight in A Room

To calculate the amount of daylight in a room, first calculate the direct light from the sky, and secondly, calculate the indirect light which consists of the reflected light from external surfaces and light received by reflection from the internal surfaces of the room.

The total of direct and indirect light gives the total daylight. This is usually expressed as the daylight factor in the room, that is, the ratio of the light in the room to the light of the unobstructed sky.

Direct daylight

The direct light from the sky which reaches any given point in a room is determined by how big a patch of sky can be seen from that point (Figure 1), or more strictly, the projected solid angle subtended by the patch of visible sky at that point. It is also determined by the brightness of the patch of sky. If the brightness of the patch of sky can be assumed to be uniform, the ratio of direct internal light to the external light from the sky is known as the sky component, and it is proportional to this projected solid angle. Formulae have been worked out to enable the sky component to be calculated.

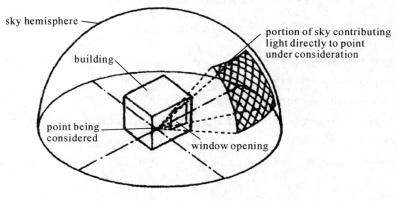

Figure 1

A simpler method of determining the direct light from the sky is by means of sky component protractors which can be laid directly on to the working drawings. Figure 2 shows the use of the protractors. An alternative method is to use published tables.

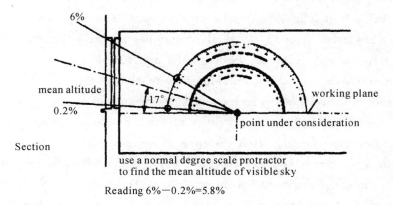

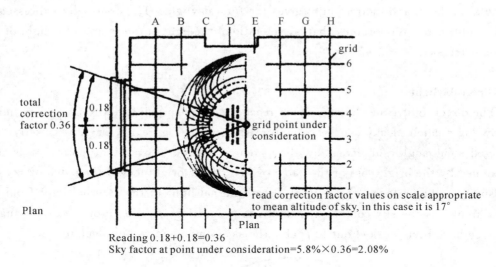

Figure 2

Answer these questions:

a) What is the purpose of determining the daylight factor?

b) What are the three sources of indirect light in a room?

c) What two factors determine the amount of direct light which reaches any given point in a room?

d) What is proportional to the area of the patch of sky seen from a given point in a room?

e) How can the sky component be determined at the design stage?

f) What is the advantage of using the protractor to calculate the sky component?

(B) Laying Bricks

Bricks are laid in the various positions for visual reasons, structural reasons or both. The simplest brick wall is a single wythe of stretchers. For walls two or more wythes thick, headers are used to bond the wythes together into a structural unit. Rowlock courses are often used for caps on garden walls and for sloping sills under windows, although such caps and sills are not durable in severe climates. Architects frequently employ soldier courses for visual emphasis in such locations as window lintels or tops of walls.

The problem of bonding multiple wythes of brick has been solved in many ways in different regions of the world, often resulting in surface patterns that are particularly pleasing to the eye. Some famous structural bonds for brickwork are Common Bond, Flemish Bond, and English Bond. On the exterior of buildings, the cavity wall, with its single outside wythe, offers the designer little excuse to use anything but Running Bond. Inside a building, safely out of the weather, one may use solid brick walls in any desired bond. For fireplaces and other very small brick constructions, however, it is often difficult to create a long enough stretch of unbroken wall to justify the use of bonded brickwork.

While conceptually simple, bricklaying requires both extreme care and considerable experience to produce a satisfactory result, especially where a number of bricklayers working side by side must produce identical work on a major structure. Yet, speed is essential to the economy of masonry construction. This level of expertise takes time and hard work to acquire. The laying of leads is relatively labor intensive. The work is checked frequently with a spirit level to assure that surfaces are flat and plumb and courses are level. When the leads have been completed, a mason's line(a heavy string) is stretched between the leads, using reshaped line blocks at each end to locate the end of the line precisely at the top of each course of bricks. The laying of the infill bricks between the leads is much faster and easier because the mason needs only a trowel in one hand and a brick in the other to lay to the line and create a perfect wall. It follows that leads are expensive as compared to the wall surfaces between, so that where economy is important the designer should seek to minimize the number of corners in a brick structure.

Mortar joints can vary in thickness from about 1/4 inch(6.5mm) to more than 1/2 inch(13mm). Thin joints work only when the bricks are identical to one another within very small tolerances and the mortar is made with fine sand. Very thick joints require a stiff mortar that is difficult to work with. Mortar joints are usually standardized at 3/8 inch(9.5mm), which is easy for the mason and allows for considerable distortion and unevenness in the bricks. One-half-inch joints are also common. The joints in brickwork are tooled 1 hour or 2 after laying as the mortar begins to harden, to give a neat appearance and to compact the mortar into a profile that meets the visual and weather-resistive requirements of the wall. Outdoors, concave joints shed water and resist freeze-thaw damage better than others. Indoors, stripped joints can be used to accentuate the pattern of bricks in the wall.

Finally, brick walls are given a final cleaning by scrubbing with acid(HCl) and rinsing with water to remove mortar stains from the faces of the bricks. Light-colored bricks can be stained by acids, and should be cleaned by other means.

New Words and Expressions:

visual   a.   视觉的
row lock n.   行锁
sill   n.   窗台,门槛
durable   a.   持久的,耐久的
lintel   n.   过梁,楣石
exterior a.   外部的
cavity   n.   洞腔
masonry   n.   砖石,砌筑

expertise	n.	专业知识,技能
plumb	n.	测深锤,铅锤
precisely	ad.	精确地
mason	n.	砖石工,泥瓦匠
trowel	n.	泥刀,抹子
mortar	n.	砂浆,灰浆
standardized	v.	使标准化
distortion	n.	变形,扭曲
profile	n.	侧面,轮廓
concave	a.	凹的
shed	n.	平房,棚
strip	v.	使从某处除去,卸掉
accentuate	v.	突出,强调
scrub	v.	刷洗,取消
rinse	v.	冲洗
acid	n.	酸

Post-Reading Exercises:

1. Multiple choices

1) In different regions of the world, often surface patterns that are particularly _____ have been developed.

   A. visible   B. obvious   C. beautiful   D. skillful

2) Bricklaying is simple _____, but requires both extreme care and considerable experience to produce a satisfactory result.

   A. in theory   B. at the beginning

   C. during daytime   D. in summer

3) It is _____ work to lay leads.

   A. a lot of   B. more   C. major   D. little

4) To lay the infill bricks, you need _____ to create a perfect wall.

   A. one hand   B. the left hand

   C. the right hand   D. both hands

5) Because leads are expensive, the designer should _____ to minimize the number of corners in a brick structure.

   A. search   B. try   C. hope   D. help

6) Mortar joints vary in thickness from about 1/4 inch(6.5mm) to more than 1/2 inch (13mm).

   A. have to   B. may   C. need to   D. are able to

7) Thin joints work only when the bricks are of the same _____ within very small tolerances and the mortar is made with fine sand.

  A. design   B. color   C. size   D. weight

  8) Light-colored bricks can be _____ by acids, and should be cleaned by other means.

  A. damaged   B. repaired   C. improved   D. colored

  2. Decide whether the following statements are true(T) or false(F) according to the passage given above:

  1) For walls more than 1 m thick, headers are used to bond the wythes together into a structural unit. ( )

  2) Headers are frequently used for special visual emphasis in such locations as window lintels or tops of walls. ( )

  3) Inside a building there are no restrictions, which kind of bond should be used. ( )

  4) Speed is essential to the ecology of masonry construction. ( )

  5) Because laying of leads takes a long time, the work has to be checked frequently. ( )

  6) It takes much more time to lay the infill bricks between the leads, because the mason needs a trowel in one hand and a brick in the other to create a perfect wall. ( )

  7) Mortar joints are usually standardized to correct probable unevenness in the bricks. ( )

  8) It is not necessary to clean brick walls with acid and rinsing with water in the end. ( )

UNIT 9 PROCESS

Part 1 Warm-up Activities

1. Assembly sequence of a prefabricated building.
The sequence is divided into four stages or phases:
Phase 1

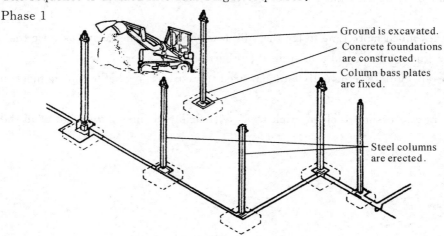

- Ground is excavated.
- Concrete foundations are constructed.
- Column bass plates are fixed.
- Steel columns are erected.

First, the ground is excavated. Then, the concrete foundations are constructed. Later, the column base plates are fixed. Finally, the steel columns are erected.

Now look at the drawings of the next three stages and make statements about the sequence of events in phases 2, 3 and 4.

Phase 2

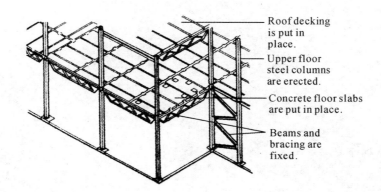

- Roof decking is put in place.
- Upper floor steel columns are erected.
- Concrete floor slabs are put in place.
- Beams and bracing are fixed.

Phase 3

Phase 4

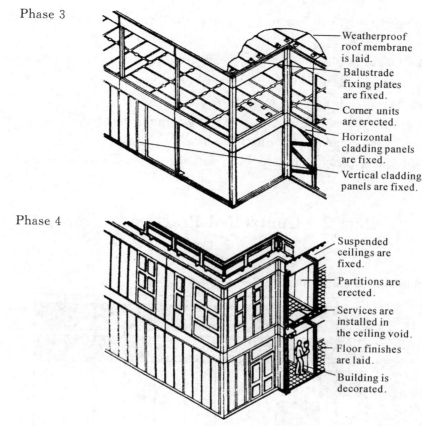

2. Read these questions:

Why are the upper floor steel columns erected *before* the roof decking has been put in place?

Why is the roof decking put in place *after* the upper floor steel columns have been erected?

Now read the answer:

Because the roof decking requires the upper floor steel columns to support it.

Look at the drawings for phases 1, 2, 3 and 4 and make similar questions to which these are the answers:

a) Because the concrete foundations require solid ground to support them.

b) Because the column base plates need a flat rigid surface to support them.

c) Because the steel columns transmit their loads through them to the foundations.

d) Because the concrete floors are supported by the beams.

e) Because the weatherproof membrane is laid over the balustrade fixing plates.

f) Because the horizontal cladding panels are fixed to the corner units.

g) Because the vertical cladding panels are fixed to the horizontal cladding panels.

h) Because the workmen require access to the ceiling void to install the services.

i) Because the partitions are fixed to the suspended ceilings.

3. Identify the part of the building or the phase of the assembly sequence described in these sentences:

a) This cannot be put in place until the upper floor steel columns have been erected.
b) Before fixing these, the workmen erect the corner units.
c) During this phase the beams and bracing are fixed.
d) The workmen fix these after constructing the concrete foundations.
e) The electric wiring is installed during this phase.
f) When the balustrade fixing plates have been fixed, the workmen can start laying this.

Part 2 Controlled Practices

1. Look at this time schedule of a building project:

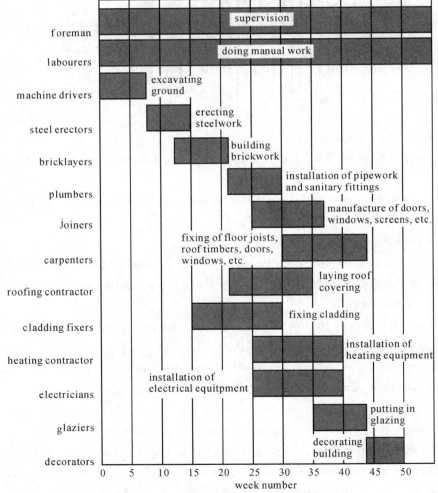

Bar chart of the sequence of trades on a building site

UNIT 9 PROCESS

2. Read the Paragraph:

During week 13, 14 and 15 (From the beginning of week 13 to the end of 15), the steel erectors work at the same time as (simultaneously with) the bricklayers. While the former erect the steelwork, the latter build the brickwork. As soon as (immediately after) the steel erectors have finished, the cladding fixers begin.

3. Use the bar chart to help you label the following drawings:

Trade: steel erectors
Job: erecting the steelwork
Weeks working: 9 to 15

Trade: cladding fixers
Job: fixing the cladding
Weeks working: 16 to 30

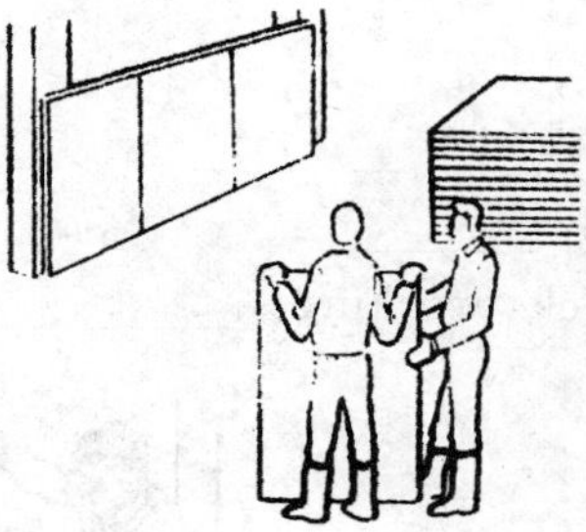

Trade: bricklayer
Job: building the brickwork
Weeks working: 13 to 21

a) Trade: _____
Job: _____
Weeks working: _____

Trade: _____
Job: _____
Weeks working: _____

Trade: _____
Job: _____
Weeks working: _____

b) Trade: _____
Job: _____
Weeks working: _____

Trade: _____
Job: _____
Weeks working: _____

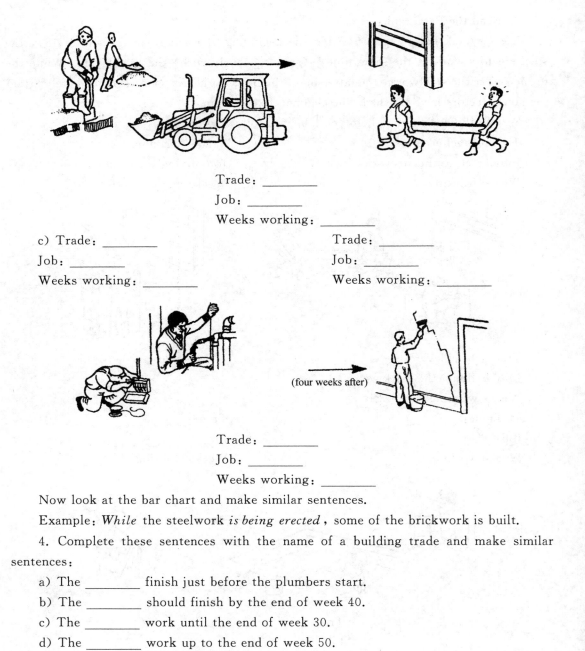

Trade: _____
Job: _____
Weeks working: _____

c) Trade: _____
Job: _____
Weeks working: _____

Trade: _____
Job: _____
Weeks working: _____

(four weeks after)

Trade: _____
Job: _____
Weeks working: _____

Now look at the bar chart and make similar sentences.

Example: *While* the steelwork *is being erected*, some of the brickwork is built.

4. Complete these sentences with the name of a building trade and make similar sentences:

a) The _____ finish just before the plumbers start.

b) The _____ should finish by the end of week 40.

c) The _____ work until the end of week 30.

d) The _____ work up to the end of week 50.

e) The _____ should finish no later than the end of week 8.

Part 3 Further Development

Read the following sentences and learn them by heart.

1) I am a manager (project manager, resident construction manager, construction superintendent, controller, staff member, engineer, technician, economist, supervisor, foreman, worker). 我是经理(项目经理、驻工地总代表、工地主任、管理员、职员、工程师、技术员、经济员、检查员、工长、工人)。

2) My technical specialty is civil engineering (chemical engineering, process, mechanical equipment, electrical, instrumentation, piping, welding, furnace building, corrosion prevention, thermal-insulation, heating-ventilation, quality control). 我的技术专业是土建工程(化工工程、工艺、机械设备、电气、仪表、管道、焊接、筑炉、防腐、保温、采暖通风、质量管理)。

3) What is your specialty? 你的专业是什么?

4) I am a mechanic (Electrician, pipelayer, welder, carpenter, turner, blacksmith, builder, erector, riveter, rigger, concrete worker, engine-driver, repair worker). 我是一名机械工(电工、管工、焊工、木工、车工、铁工、建筑工人、安装工人、铆工、起重工、混凝土工、司机、维修工)。

5) Our company cover all construction activities, that is: piling, civil engineering, mechanical erection, piping, electrical, instrumentation, painting and insulation work. 我们公司涉及所有施工活动,包括:打桩、土建工程、机械安装、配管、电气、仪表、油漆和保温绝缘工作。

6) What is the feature of this cracker (cracking furnace, heating furnace, reactor, mixer, centrifuge, belt-conveyer)? 这台裂解器(裂解炉、加热炉、反应器、搅拌器、离心机、皮带输送机)的特点是什么?

7) The cooler (condenser, separator, boiler, generator, scrubber, stripper, heat exchanger) is a pressure vessel. It is subject to the pressure vessel code. 这台冷却器是一个压力容器,它必须服从压力容器法规(冷凝器、分离器、锅炉、发生器、洗涤器、汽提器、热交换器)。

8) The pressure vessel must be inspected by our authoritative organization-Administration of Labor. 压力容器必须接受我们的权威机构劳动总局的监察。

9) This low (middle, high) pressure blower (pump) will be assembled in the No. 3 workshop. 这台低(中、高)压鼓风机(泵)将在三号车间里予以装配。

10) We are adjusting (installing, checking, aligning, leveling, purging) the equipment. 我们正在调整(安装、检查、找正、找平、清洗)这台设备。

11) The working team will finish the job next week. 工作班组将在下周干完这活。

12) We can adjust the levelness of the machine by means of shim and screw jack. 我们

可以利用垫铁和螺丝千斤顶来调整机器的水平度。

13) After seven days, the grouted mortar will have concreted, and then we shall tighten the anchor bolts. 灌浆在七天以后凝固,我们就将拧紧地脚螺栓。

14) How many radial(axial) clearance are there in this bush(journal bearing, thrust bearing)? 这个轴套(轴颈轴承、止推轴承)的径向(轴向)间隙是多少?

15) We prefer welding to riveting. 我们认为焊接比铆接好。

16) Do you know how to assemble(adjust) this new machine? 你知道如何装配(调整)这台新机器吗?

17) To maintain the best quality of the construction work is the important responsibility of the field controllers. 保持施工工作的优良质量是现场管理人员的重要职责。

18) I want to see the certificate of quality(certificate of manufacturer, certificate of inspection, certificate of shipment, material certificate, certificate of proof). 我要看看质量证书(制造厂证书、检查证明书、出口许可证书、材料合格证、检验证书)。

19) We have received Certificate of Authorization for the fabrication and erection of pressure vessels. 我们具有压力容器制作和安装的授权认可证书。

20) Check list(quality specification) has been signed by the controller(inspector, checker). 检验单(质量说明书)已由管理员(检查员、审核人)签字。

Part 4　Business Activities

1. 文化与指南(Culture and Directions)

当地劳务管理(management of local labor)是国际工程管理的又一个方面。由于工程所在国往往对承包商带入该国的工程人员的数量有所限制,或由于承包商雇用本国劳动力费用较高,所以,承包商一般只从本国选派一些关键技术人员和管理人员(key personnel),而全部普工(unskilled)、半熟练工人(semi-skilled)和一些技工(skilled)均从工程所在国雇用。

招工时,要让求职人出示法律规定的必要证件,如身份证(ID card)、出生证(certificate of birth)等。之后,可询问其工作经历(working experience)。对于技术工人,还应进行技术测试,并规定试用期(probationary period),以避免因招收的人员技术欠佳或缺乏责任心而导致设备的坏损。

承包商的劳务管理人员一定要熟悉当地的劳动法的有关规定,如:最低工资标准;节假日加班费的标准;工人的福利;劳动保护;雇用外籍劳工的政策;人身保险等。劳动部门是代表工人的利益来监督管理雇主的。与当地劳动部门搞好关系将有助于承包商的劳务管理工作。

2. 情景会话(Situational Conversations)

(1) Hiring Local Labor

(For some reasons, the contractor cannot bring sufficient manpower to the project owner's country and, therefore, has to employ local labour. A lot must be done in order to have a highly productive local workforce. Now several equipment operators and dump truck drivers are needed for the power project. After the job vacancies have been advertised in the newspaper, the local job seekers come to the site office for interviews. The contractor's labor officer is receiving them.)

(A: A labor officer from the contractor; B: The job seekers)

B: Excuse me, I saw in your advertisement that you have some vacancies for dump truck drivers. Are they still available?

A: We still have two vacancies for ten-wheel dump truck drivers.

B: Lucky enough. I am looking for a job as a driver.

A: May I see your driving license?

B: Certainly, here it is.

A: How long have you been driving this kind of truck?

B: Ten years. Here is a letter of recommendation from my previous employer.

A: Fine. May I ask why you quitted the job there?

B: I live in San Ignacio. The company I worked with is too far from my home. Besides, the transportation is not convenient.

A: Do you have your ID card with you?

B: Yes. Here you are.

A: And your Social Security card, please.

B: I had one, but I lost it last week. If you accept me, I will apply to the Social Security Board for a new one. I think I can get it within a week.

A: OK. This is an application form. Please fill it in. Tomorrow you come here for a driving test. If you pass it, you will be a probationary worker for two weeks. That means we will try you for two weeks. If you prove to be a qualified driver during that period, you will become a formal worker and can work with us until this hydro project is completed.

B: How much do you pay, please?

A: The starting wage is 25 dollars for a day shift and 30 for a night shift. One shift is seven hours and a half, with extra pay for overtime work, if any. You will have the chance to get a rise if you prove to be a good worker.

B: Do I have to live on the site camp? If not, how do I get to the job site everyday?

A: We provide free transportation for all our workers. You said you live in the suburb? Most of our local workers come from this area. We have a bus stop there directly to the job site.

B: That's good. So what time do I come here for the test tomorrow?

A: You can come on our bus which leaves the suburb at six thirty in the morning. When you arrive, please come to me. I'll arrange the test for you.

B: All right. What if the bus driver won't let me get on the bus?

A: Don't worry. Take this note and show it to the driver.

B: Thank you, sir, I'll see you tomorrow.

A: OK, see you tomorrow and good luck.

(2) A Meeting with the Labor Office

(The contractor's representative and the owner's representative are having a meeting on labor issues with a labor officer at the office of the Labor Department)

(A: the owner's representative; B: the contractor's representative; C: a labor officer of the Labor department)

C: Gentlemen, I've asked you to be here today to have a talk on some labor issues. Recently, my department has received complaints from workers employed at the project. They've complained that they're often underpaid and that the working conditions are poor. I'd like to know the roles played by the owner and the contractor.

A: Let me briefly explain the responsibilities of each side. As the owner, we are the investor of the project and finance the project only. It is the contractor that hires the workers and constructs this project directly. We are required by the contract to give them assistance in this respect. Therefore, they are responsible for paying the workers and taking care of the benefits.

B: That's true. At present, we do have some problems with the local workers. One reason is the poor communications between the Chinese foremen and the local workers because of the language barrier. We are giving the Chinese foremen a short English language training program so that they can talk to the local workers in simple English to avoid misunderstandings. I have another point to make about the local workers' pay. According to relevant laws of your country, as the employer of the workers here, we are responsible for deducting their Social Security contributions and income tax from their weekly pay and submitting them to the Social Security Board and the Income Tax Office respectively. Therefore, on the payroll, wages are not as much as they thought they should get.

C: Yes, it's legal to deduct their Social Security contributions and income tax, Mr. Zhang. However, before you make any deductions, you should make it clear to all the workers, and as our local practice, all the deductions must be shown on the payrolls so that the workers are clear about how much their take-home pay should be. You know, the workers are quite conscious of their pay because most of them have a family to support. If they think they are underpaid, it will affect their turnout and their productivity.

B: I quite agree with you. I'll adopt your suggestions.

C: How about the working conditions for all the workers?

B: We have provided hardhats and raincoats to all the workers free of charge. We also provide free transportation.

C: That's not enough. For such a project, canvas gloves and proper shoes are also required for work protection under our labour law.

B: OK. We'll do that.

C: Is drinking water available at the major work areas?

B: We have ordered two pickups from the United States, especially for delivery of drinking water to each work area. As soon as they are here, we will put them in use.

C: What are the working hours and pay rates?

B: We now have two shifts: 7 a.m. —3 p.m. and 3 p.m. —11p.m. Each shift has a half-hour lunch break. The short lunch break has been requested by the workers so as to shorten their time at the site and go back home earlier.

C: Let me tell you, Mr. Zhang and Mr. Carson, that Section 121 of the Labor Law requires a one-hour lunch break unless approval has been given by the Commissioner of Labor.

B: Will you get such approval from the Commissioner for us, Mr. Carson?

A: All right, but give me a letter of application before I seek it.

B: Will you please get me a copy of the Labor Law, Mr. Young, so that I can get a complete understanding of the stipulations?

C: Here is one copy of the condensed version for you.

B: Thank you.

C: I have to remind both of you that all the workers are entitled to a two-week vacation with pay for every twelve months of work. They're also entitled to sixteen days of sick leave with pay after sixty days of work if they provide valid medical evidence. One thing to which I'd draw your particular attention: If the workers are required to work on public holidays, make sure that they get reasonable pay. You can find the way to calculate how much the workers should be paid in the Labor Law.

A: As the project owner, we will urge the contractor to follow all relevant regulations.

B: We will try our best to make our workers satisfied with us.

C: Let me reaffirm that every employer must comply with the Labor Law and pay serious attention to all the workers' welfare. From my point of view, a happy

workforce in a good working environment is most productive.
B: I certainly agree.

New Words and Expressions:

ten-wheel dump truck	十轮自卸车
Social Security card	社会保险卡
probationary worker	试用工
free transportation	免费交通
labor issues	劳工问题
labor officer	（当地）主管劳动就业的官员
Labor Department	（当地）主管劳动就业的部门
take care of the benefits	负担津贴
payroll n.	工资单
make any deductions	扣除
take-home pay	实得工资
be quite conscious of	对……很敏感（在意）
turnout n.	出勤，产额
canvas gloves	帆布手套
commissioner	专员，政府部门的负责人
stipulation n.	规定
condensed version	压缩版本
vacation with pay	带资休假
sick leave with pay	带资休病假
valid medical evidence	有效的医疗证明
workers' welfare	工人的福利
comply with the Labor Law	遵守劳动法

Notes to the Text:

1. local labor:当地劳工。labor 为"劳工"的总称，不可数，若说"十个工人"，则为 ten laborers.

2. job vacancy:招工名额。vacancy 的意思是"空缺"，其形容词形式是 vacant，如到一个旅馆(hotel)要住宿，常问:Do you have any vacant rooms?（还有空房间吗？）

3. Social Security Board:社会保障局。在某些国家，法律规定雇主所雇人员必持有社会保障局的保险卡(Social security card)。若没有，雇主自己或雇员本人必须限期为其雇员到 Social Security Board 办理所规定的险别，保险费(social security contributions)由雇主和雇员按一定的比例分担，保险费的标准按雇员的工资多少而定，雇主可从雇员的工资中扣除其应交份额(share)，连同雇主应付之份额，定期（一般为一个月）向社会保障局缴纳。如果雇员在受雇期间生病、在工作时出现受伤或死亡等情况，雇主只需负担相当于该雇员几天工资的补偿费(compensation)，其余费用均由社会保障局负担。倘若雇员没有社会保险卡而为雇主工作,或雇主没有按期缴纳保险费,则雇主可能遭到严厉的惩罚(severe penalty)。

    4. try you for two weeks:试工两周。

    5. starting wage:起始工资。

    6. day shift:白班。shift 在这里是"轮班工作"的意思,如:morning shift(早班),evening shift(晚班),night shift(夜班)。

    7. foreman:工长。英语类似的表达还有 leading hand(领班)。

    8. social security contributions:社会保险费。

    9. income tax:所得税。"个人所得税"是 personal income tax。

    10. local practice:当地惯例。又如:international practice(国际惯例)。

    11. hardhat:安全帽。也可以说 safety helmet。

    12. under our local law:按我们当地的法律。

    13. pay rate:工资标准,指每小时或每天的工资。

Post-Reading Exercises:

Now try to do the following translations:

1. "I had one, but I lost it last week. If you accept me, I will apply to the Social Security Board for a new one。

"apply (to...) for..." is used frequently in construction management, "(向……)申请……"

    1) A: Can we employ foreign workers?

    B:_____

(一般情况下可以,但你们须向劳动司为他们申请劳动许可证,并说明理由。)

    2) A: We have decided to import your table lights to our country.

    B:_____

(谢谢,我希望你们能协助我们在你们国家为我们的商标申请注册。)

    3) A:_____

(请你方用快件将所用单据寄给我们,以便我们能够申请进口许可证。)

    B: We'll do it in a couple of days.

2. "What if the bus driver won't let me get on the bus?"(如果司机不让我上车怎么办?)

"What if...?" is used often when the speaker is in doubt about something and needs clarification.

    1) A: Maria, please book me a ticket on a direct flight to Seattle.

    B:_____

(如果没有直达班机是否还预订?)

    2) A:_____

(如果明天下雨怎么办?)

    B: Don't stop pouring the concrete if it's not heavy rain.

3) A: _____

（钢筋生锈了怎么办？）

B: They will be rejected by the resident engineer.

3. "One thing to which I'd like to draw your attention."（有一件事我想提醒你注意。）

"draw one's attention to sth." is used often when you want someone to take notice of something.

1) A: There's no stipulation about it in the specification.

B: _____

（我想提醒你看一看 5 月 2 日你方下达的变更命令。）

2) A: We have informed you of our intention to change the design already.

B: _____

（请你看一看我们 6 月 8 日对你方的答复好吗？）

3) A: _____

（虽然是最后一点，但也很重要，我想提醒你方注意这一里程碑日期。）

B: We'll certainly keep it in mind.

Part 5　Extensive Reading

(A) Site Labor

The two most important men on a construction site are the contractor's agent and the navy. The agent is important because he acts for the contractor and has the authority of the contractor on the site. The navy is important because he, with perhaps hundreds of other navies, does most of the work. Compared with these two, although no civil engineer will happily admit it, the resident engineer comes third, for without him the work would be completed in time, though perhaps badly and at high cost to the client. Most of the civil engineering design has been done by the time the contract has been signed, and the contractor is bound by to follow the design as it is laid down in the drawings, though normally no contract is left without the supervision of a resident engineer or at least of a clerk of the works or a civil engineering inspector.

Between the agent and the navy there are many other men who organize the work, help it to go smoothly, see to the arrival of the essential materials, check their quality and help to get the contract finished on time. The chief of these is the general foreman, a man who has worked for many years on construction sites, either as a navy and then as a ganger in charge of a group(gang) of navies, or more usually, as a tradesman and then a foreman of his trade.

"Tradesman" in civil engineering or building is the name given to masons, bricklayers, carpenters, plasterers, mechanical fitters and others who do a special kind of work with their hands(a trade).

The critical trade in civil engineering is usually carpentry because the carpenters build the concrete formwork, and the foreman carpenter sets out the formwork for them and often also sets out for the other trades. If the setting out is not well ahead of its time, the concrete formwork is likely to be delayed, with serious consequences for the progress of the job. Setting out means marking out the work on the ground or on the structure, to ensure that it is properly placed. A civil engineer usually sets out the main location of the building, and ties it in with his surveying instruments to local survey monuments.

Since navies do the concreting and digging which are essential to most foundation work, and foundations are the first work after the demolition and site clearance, navies are usually the first men employed on a building site.

There is a natural sequence of construction which is the same for most buildings, and for this reason, the trades come one to a site usually in the same sequence, from one structure to the next. This sequence is shown by the order in which the trades are printed in the bill of quantities. It is important to adopt one particular sequence because in that way the writer of the bill can be sure that he has forgotten nothing. There are (1) demolition and shoring, (2) earthwork, (3) piling, (4) concreting, (5) precast concrete and hollow floors, (6) brickwork and partitions, (7) drainage and sewage, (8) asphalt, (9) paving, (10) masonry, (11) roofing, (12) timber and hardware, (13) structural steelwork, (14) metalwork, (15) plastering, wall tiling and terrazzo, (16) sheet metal, (17) rainwater gutters and down-pipes, (18) cold-water supply and-sanitary plumbing, (19) hot-water supply, (20) gas and water mains, (21) heating, (22) ventilating, (23) electrical, (24) glazing, (25) painting, (26) specialist work.

New Words and Expressions:

to act for…	代表……办事
to be bound by	理应,必须
with serious consequences	带来严重的后果
site n.	工地,场所
contractor n.	合同者,承包者
navy n.	壮工
authority n.	政权,授权,权威
resident n.	居住的,居民
client n.	工程委托人,甲方代表
bind n.	捆,扎,约束
lay down	施工
drawing n.	图样,素描
supervision n.	监督,管理

inspector	n.	检查者,监察者
trades man	n.	手艺人,工匠
foreman	n.	工头
mason	n.	石工
bricklayer	n.	砖工,瓦工
carpenter	n.	木工
plasterer	n.	抹灰工
mechanical	a.	机械的
fitter	n.	装配工,钳工
form	n.	模板
demolition	n./v.	拆迁
shoring	n.	支撑,加固
pilling	n.	打桩,堆积
concreting	n.	浇灌混凝土
brickwork	n.	砖房,砖砌(体)
partition	n.	隔墙,隔板,隔开,分隔
drainage	n.	排水,排水设备,排水法
sewage	n.	污水处理
asphalt	n.	沥青,柏油
paving	n.	铺路,铺平
hardware	n.	五金,金属制零件
metalwork	n.	金属制造
plastering	n.	抹灰
terrazzo	n.	水磨石
gutter	n.	雨水槽
sanitary	a.	卫生的,关于环境卫生的
plumbing	n.	管子工程
heating	n.	加热,供热
ventilating	n.	通风
glazing	n.	油漆
painting	n.	上漆
set out		放线

Post-Reading Exercises:

Translate the following expressions:

1) site labor

2) a construction site

3) to have the authority of the contractor

4) at high cost to the client

5) set out

6) get the contract finished on time

7) a foreman of his trade

8) set out the main location of the building

9) build the concrete form work

10) adopt one particular sequence

11) 承包商行

12) 承包商代理人

13) 驻地工程师

14) 给其他工种放线

15) 土木工程检查员

16) 经管主要材料到货情况

17) 给木工定模板尺寸

18) 确保一事无遗

19) 工程进展

20) 按图纸规定的要求施工

(B) Manpower Planning For Project

Site labor must be actively managed. Planning a construction project means not only deciding what is to be done, but who will do it and when.

The manager has to forecast project manpower requirements, taking into account the availability of various kinds of labor(including sub-contract labor). They need to avoid sharp fluctuations in manning levels and the overall resource pattern for the project.

Even when unemployment is high, certain types of labor may be in short supply and the firm may have to train people to meet its needs. A shortage in just one trade or occupation may make nonsense of a contract program and add considerably to the project duration. Recognizing such problems at the outset can lead to better balancing of work and reduce peak manpower needs. Site labor should be built up and run down in a planned way, avoiding sudden changes in the numbers of each trade on site. This means scheduling project activities so that cumulative needs do not exceed the labor available and do not fluctuate too much. Manpower planning will not guarantee high productivity or good labor relations, but its absence can lead to poor performance and strained relations.

A manning curve is a useful way to summarize labor requirements and highlight problems. It is compiled by combining the labor requirements for each trade. The chart can be used to balance peak demand and to record numbers of employees used to date. It shows the rate of build-up or run-down of labor and any correction needed to keep on schedule. In the curve, the information is obtained from a network analysis and shows manning curves' based on the earliest and latest starts for each activity on the program. The manpower build-up for the earliest start curve is high. It emphasizes that when too many people are taken on at once, supervisors become overloaded and productivity can be difficult to sustain.

Too steep a descent at the end of the project also causes problems. Contractors often do this to recover from delays, but the usual result is to overshoot the completion date, with labor remaining on site, tidying up loose ends and winding down long after work should have finished. Properly updated manning curves can show up early signs of under-manning, indicating the need for a review of the contract program or a change in recruiting procedure, But they cannot easily be used to show the effects of variations or additional work and are only as accurate as the estimates of labor outputs and labor availability on which they are based.

The number of employees that can work simultaneously on a project without productivity falling is limited, but there should not be too much reliance on overtime to make up for labor shortages. Regular overtime is usually expensive and output during overtime working is often lower than that achieved during normal hours.

Manpower planning may be looked at collectively or individually. Manning requirements for main trades are normally assessed collectively, taking account of quality of work and output standards. Technical and administrative tasks are less easy to quantify in man hours. It is often necessary to individually assess staff, such as managers, engineers and surveyors. A useful check on what to allow can be obtained from past records of ratios of indirect to operative labor.

New Words and Expressions:

fluctuation	n.	起伏,波动
fluctuate	v.	给……配备员工
outset	n.	开端,开始
peak	a.	最高的,高峰的
cumulative	a.	渐增的,累加的
guarantee	v.	保证,担保
strained	a.	紧张的,勉强的
curve	n.	曲线,曲线图
compile	v.	编辑,汇编
chart	n.	图,图表
run-down	n.	裁减,裁员
sustain	v.	支撑,经受住
steep	a.	急转直下的
descent	n.	下降,降下
overshoot	v.	超过,使(事情)做过头
simultaneously	ad.	同时地
man-hour	n.	工时

surveyor *n.*	检查员
at/in a plan	在开头时
build up	增长,集结
tidy up	整理,收拾
wind-down	进展缓慢
earliest start manpower	劳力最早进入数

Post-Reading Exercises:

Decide whether the following statements are true(T) or false(F) according to the passage given above.

1) Planning a construction project means deciding what is to be done.　　　　(　)

2) The manager has to forecast project manpower requirements and also all the probabilities about it.　　　　(　)

3) A shortage in just one trade or occupation may result in a failure of a project.
　　　　(　)

4) Manpower planning will guarantee high productivity or good labor relations.
　　　　(　)

5) Too steep a descent at the end of the project is a good way for contractors to recover from delays.　　　　(　)

6) A manning curve is a useful way to summarize labor requirements and highlight problems.　　　　(　)

7) Output during overtime working is often lower than that achieved during normal hours.　　　　(　)

UNIT 10 FUNCTIONS OF BUILDING

Part 1 Warm-up Activities

1. Match the pictures with the names of the building and work in groups to discuss what each building is used for and what activities will go on in it.

| block of flats | fire station | railway station | hotel |
| school | swimming pool | hospital | church | bank |

UNIT 10   FUNCTIONS OF BUILDING

2. Look at this table and make Q-A (question and answer) as the examples below:

Building type	Purpose of building	Examples of activities	Spaces provided
university	educating 2000 students per year	giving lectures a) _____ storing, reading books	lecture room laboratory b) _____
house	accommodating a family of 5 persons	Preparing food c) _____ d) _____	e) _____ dining room bedroom
hospital	treating 150 patients per day	f) _____ g) _____	treatment room emergency room
factory	making 400 precast concrete panels per week	storing materials h) _____	i) _____ j) _____

Examples:

Q: What is the *function* of a university?

A: A university functions/serves as a place for educating students.

Q: What spaces *are provided* in the building?

A: Spaces *provided* in the building include a lecture room, laboratories and a library.

Q: What *is* the lecture room *used* for?

A: The lecture room *is used for* giving lectures.

3. Buildings are designed so they are capable of performing the design requirements. The most important design requirements include the following; please match the design requirements with its correspondent functions.

weather resistance   to resist loads
privacy   to prevent fire from spreading
surfaces   to provide fresh air
security   to provide visual screening
fire resistance   to keep out intruders

structure	to keep out wind, dust and precipitation
ventilation	to provide surfaces for activities
thermal insulation	to control sound transmission
sound insulation	to modify the passage of heat
moisture	to provide natural and artificial light
light	to control the passage of moisture

Part 2   Controlled Practices

1. Match the building types on the left with the phrases on the right to make similar sentences as the example.

Example: *The university has the capacity to educate/is capable of educating 2,000 students a year.*

university	educate 2,000 students a year
house	make 400 precast concrete panels per week
hospital	accommodate a family of five persons
factory	serve up to 200 customers per day
post office	treat up to 150 patients per day
shop	handle up to 1,000 letters per day
railway station	deal with 10 train movements per day

2. Look and read:

Buildings are designed in such a way they are capable of meeting several requirements. The most important design requirements include the following:

A) weather resistance—keep out wind, dust and precipitation

B) privacy—provide visual screening

C) surfaces—provide surfaces for activities

D) security—keep out intruders

E) fire resistance—prevent fire from spreading

F) structure—resist loads

G) ventilation—provide fresh air

H) thermal insulation—modify the passage of heat

I) sound insulation—control sound transmission

J) moisture—control the passage of moisture

K) light—provide natural and artificial light

UNIT 10 FUNCTIONS OF BUILDING

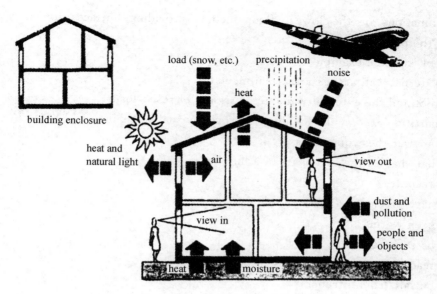

External envelope

Internal division

3. Complete the table by choosing from letters A to K above to indicate the requirements.

	Element	Main functions
External envelope	lowest floor	C, H, J, etc.
	external wall	
	roof	
Internal division	suspended floor	
	partitions	
	suspended ceiling	

Use the table to make statements like the following:

149

The functions of the lowest floor include providing surfaces for activities and modifying the passage of heat.

4. Look at these examples:

The external wall *acts as* a thermal insulator.

The roof and the external walls *are designed to* resist loads.

The partition *enables* the building to provide visual screening.

Now answer these questions:

a) What enables the occupants of a building

—to keep dry?

—to have privacy?

—to keep warm?

—to be safe from fire?

—to read during the nighttime?

—to be safe from intruders?

b) What element is designed to

—control the noise level between rooms?

—support snow loads?

—resist the passage of moisture?

—let in natural light?

—control the movement of people into and out of the building?

c) What elements act as

—a thermal insulator?

—a sound insulator?

—a space divider?

—a filter to separate the internal volume from the external environment?

—a moisture barrier?

5. Make statements about the properties of the materials given in the table below.

Example: *Concrete is capable of withstanding high temperatures.*

Name	Properties
concrete	low combustibility, high density, pervious
aluminium	impervious, corrosion resistant, high strength
steel	high strength, high thermal conductivity
mineral wool	low thermal conductivity, low strength
ceramic tile	hard, impervious, good appearance

6. Look at these sections through walls and complete the labels:

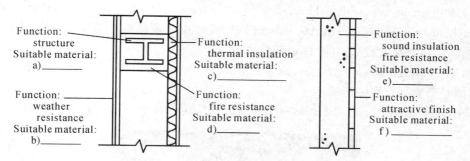

Now write five sentences like this example.

Concrete has low combustibility and is therefore used to provide fire resistance for walls.

1) _____.
2) _____.
3) _____.
4) _____.
5) _____.

Part 3 Further Development

About Architecture Schools(建筑风格流派)

1. Classical style(BC800—AD1550)古典主义风格

Acropolis 雅典卫城

1) Ancient Greece architecture(BC800—AD300)古希腊建筑风格

(1) Central is a living room, hall, surrounded by columns, can be collectively referred

to as central pillar construction.（中央是厅堂,大殿,周围是柱子,可统称为环柱式建筑。）

（2）Column of the style. There are four styles：①Doric column，②Ionic column，③Corinthian column，④girl statue column.（柱式的式样共有四种：①陶立克柱式,②爱奥尼克柱式,③科林斯式柱式,④女郎雕像柱式。）

（3）Greek architecture and Greek sculpture is tightly bound together.（希腊的建筑与希腊雕刻是紧紧结合在一起的。）

2）Roman architecture（BC300—AD365）古罗马建筑风格

Roman Colosseum　古罗马斗兽场

Ancient Roman architecture bearing the ancient Greek civilization's architectural style is also a development of ancient Greek architecture.（古代罗马建筑承载了古希腊文明中的建筑风格,同时又是古希腊建筑的一种发展。）

3）European medieval architecture（AD400—AD1400）欧洲中世纪建筑风格

Cologne Cathedral　科隆主教堂

Romanesque church looks like a feudal lord's castle, the cross plane, the short horizontal, vertical length, the intersection near the eastern end. This is called a Latin cross to a symbol of the cross of Christ crucified, but also to enhance the significance of religion.（罗马式教堂的外形像封建领主的城堡,为十字形平面,横向短,竖向长,交点靠近东端。这叫做拉丁十字架,以象征耶稣钉死的十字架,更加强了宗教的意义。）

First, its size and height create a new record, followed by the body very strong upward momentum, brisk vertical line over the body.（首先它在尺度和高度上创造了新纪录,其次是形体向上的动势十分强烈,轻灵的垂直线遍布全身。）

4）Renaissance style(AD1420—AD1550)文艺复兴建筑风格

Florence Cathedral　佛罗伦萨大教堂

The most obvious features of Renaissance architecture is discarded the medieval Gothic style, and in the religious and secular buildings of ancient Greece to re-use of the composition elements of Roman column. Even the architectural style of the various regions fused together with the classical column.（文艺复兴建筑最明显的特征是扬弃了中世纪时期的哥特式建筑风格,而在宗教和世俗建筑上重新采用古希腊罗马时期的柱式构图要素,甚至将各个地区的建筑风格同古典柱式融合在一起。）

2. Neo-classical architectural style(AD1750—AD1880)新古典主义风格

Simplicity: Clear and uncomplicated geometry with the use of right angles and straight lines, symmetry of form and minimal ornamentation.（新古典主义风格是简洁明快,清晰简单的几何图形以及合适的角度与直线的运用,体现形式的对称和最小的装饰。）

University of Virginia　弗吉尼亚大学

3. Modern architecture style(AD1960—)现代建筑风格

United Nations Headquarters　联合国大厦

Modern architecture pays attention to the functional and modern aspects, emphasizing the rational style created by rejecting the tradition of international, national, regional and personality to form a size-fits-building style.(现代建筑强调功能性和理性,排斥传统、民族性、地域性和个性,形成了千篇一律的建筑样式。)

4. Post-modern style(AD1980—)后现代建筑风格

The post-modern architecture is the building after the general term for all schools, raised by U.S. architect Stern; postmodern architecture has three characteristics—the use of decoration, being symbolic or metaphorical and integration with the existing environment.(后现代建筑是指现代以后的各流派的建筑总称,美国建筑师斯特恩提出:后现代主义建筑有三个特征:采用装饰;具有象征性或隐喻性;与现有环境融合。)

Portland City Hall　波特兰市政厅　　　　　　Sydney Opera House　悉尼歌剧院

Part 4　Business Activities

Safety Patrol

1. 文化与指南(Culture and Directions)

　　根据建筑施工的特点,科学、合理地制定安全检查(safety patrol)内容是搞好建筑施工安全检查的基础和关键,就当前事故发生情况和建筑施工技术手段来看,可将建筑施工安全检查分为:安全管理、外脚手架、"三宝"(安全帽、安全网和安全带)及"四口"(楼梯口、电梯口、预留洞口和通道口)防护、施工用电、龙门架与井字架、塔吊、施工机具等七个方面,这些方面基本上能够反映建筑施工安全检查的全貌,依据这七个方面进行科学合理、针对性强、可操作性强的安全检查是保障安全检查顺利进行,进而预防事故发生的前提和基础。

2. 情景会话(Situational Conversations)

(1) The Organization for Safety

(*Regular safety patrol is an effective measure to prevent risks and accidents. The client's consultant engineer, accompanied by the contractor's project manager and other managers in each section, is visiting the construction site and doing his routine monthly patrol.*)

(A: *The client's consultant engineer*; B: *The contractor's project engineer*; C: *the workers*)

A: It is the first time for us do the monthly safety patrol today. Is everybody responsible for safety from each section present?

B: Yes. We understand the importance of the safety for all site works. Nobody neglects it.

A: Up to now, all documents regarding safety requirements and standards from the employer have been issued. We hope that these principle documents will be strictly observed by everyone involved in the works.

B: OK. To start with, let me report to you the organization for safety works. From the beginning of this project, we have appointed a safety officer, who is qualified, experienced and responsible for the safety issues in our site management. He has been given sufficient time, authority and responsibility to ensure the safety of the project.

A: What is the duty of the safety officer?

B: It mainly includes several aspects. The first aspect is to give regular inspection of the site safety in association with the works supervision system, to ensure safe methods of work are being used, and all safety and health requirements are being observed.

A: Is the safety officer authorized to rectify any faults on site?

B: Yes, he has the right and power to take effective actions to rectify any faults against safety. The safety officer's second duty is to make detailed investigations to determine the cause of accidents and dangerous occurrences, and recommend possible solutions to prevent recurrences of similar accidents.

A: Would you show me the actual position of the safety officer in your general organization chart? Is it similar to a team leader or a department leader?

B: The safety officer is an important department leader. He is authorized by the project manager from issuing safety instructions for warning, giving penalties, up to firing any person who does something seriously against safety regulations. Of course, penalty actions are the special cases we have to take. Usually we give workers safety and health training. This is the third duty of the safety officer.

A: How many people in your organization are responsible for safety work?

B: There are totally ten persons full time working for safe matters. Each of them is responsible for each working section which has around 150 to 200 workers.

A: Only one man is in charge of safety matters for a section with 200 workers?

B: No. Beside section safety officer, we have a worker in supervision of safety matters in each workers group and other 5 group members during their daily operation. They are called the safety monitors and being paid additional bonus if the works is smoothly going on without any safety problem.

A: Setting up safety organization is really important and it can provide on-site safety education as well. By the way, do you have any safety training classes?

B: Yes, we have a safety training program for our staff and workers.

A: Could you give me a brief introduction to your complete training program for safety?

B: Our Safety Training Program is carried out on the basis of the Chinese Construction Ministry Standard JGJ59—88. It can be divided into two parts: part 1 is for training workers and part 2 is for training foremen and supervisors. We have been holding these two training classes since the setting up of the site.

A: What is the training class about?

B: The training contents include many aspects such as jobsite safety policy, employee responsibility under China Labor Law, company's safety regulations, construction accidents and reporting, electrical safety, personal safety equipment, crane safety, respiratory protection, fire protection and prevention, toxic substances, first aid and emergency aid procedure, and so on.

A: Then, what about the foreman and supervisors in their safety training?

B: Some training contents are similar to that of the workers, but it covers a wider range of topics such as general safety policy, supervision for safety, safety meetings, accident investigation, safety audits, and so on.

(2) Work Area Safety Inspection

(A: *The client's consultant engineer*; B: *The contractor's project engineer*; C_1, C_2: *the workers*)

A: Now let's go to the work area for a safety inspection. We will see if the safety regulations are strictly implemented. Please record all the default events or negligence of safety in each section and to make a safety mark on them in accordance with these forms. We will have a judgment summary after the inspection.

B: That's what we usually do. Now everybody, take on your helmet, please.

A: Let's start with the helmets and other personal protective equipments. (Looking around) Yes, it is good that every body on site wears a helmet. But I find that not everybody has his protective footwear, or what's even worse is that I notice some workers without any shoes. Have you given workers safety shoes?

B: Yes, we have. But since it is very hot here, local people feel uncomfortable with shoes. They often throw away shoes, even though they know these shoes are useful in protection of their feet.

A: Yes, I see, but I'm afraid this can not justify working without protecting feet. Strong measures must be taken to ensure everybody wears safety shoes. Besides, people working in dust areas, such as in cement warehouse, must wear the respirators. Such kind of things should be disciplined.

B: OK. It is really necessary to take strong actions. We can assure you that such things will not happen in the next safety inspection patrol.

A: I believe the situation will be greatly improved with our joint efforts, and I must say, I am fairly satisfied with the temporary electrical installations. All site electrical equipments including cables, transformers, and switch panels are suitably installed in a safe manner. I think the several grades of the timing switches are very useful for avoiding electrical short circuit.

B: Yes. Since the electrical tools are very frequently used in the construction site now, we must pay enough attention to avoid short circuit. We use different timing switches by 0.01, 0.03 and $0 \sim 1$ second of breaking time for different scope of electrical distribution installed in each working section. In case any short circuit happens, the relevant switch will cut off electricity immediately, so people working with electrical tools are much safer than before.

A: Oh. Look, (pointing to a fuse box) here is a wrong connection. There is a copper wire instead of the fuse. It should be modified immediately and the worker who made such mistake should be warned seriously.

B: OK. We will instruct all sections to have a thorough inspection for all fuse boxes.

A: Now I am afraid I must point out some inadequacy in your fire prevention and fire extinguishing equipment.

B: All criticisms are welcome and they are valuable to the improvement of our work.

A: In the steel workshop, the empty and charged gas cylinders are not stored in separate locations and the empty cylinders are not marked identifiably. In some areas the welding are operated above the ground, but some combustible and flammable materials still remain. They are very likely to cause fire accident.

B: We will take immediate actions to separate the gas cylinders in different rooms with conspicuous labels and places for welding operation will be provided with fire extinguishers.

A: And, it is necessary to provide several container made by non-combustible materials for people to smoke inside only. In this way, smoking ban in the working areas can be effectively practiced.

B: But it will cost too much. How about providing certain open areas for people to smoke?

A: It may be economical, but how to keep the open areas safe is also a big problem. There are potential safety hazards unless they are enclosed to avoid any sparks. You'd better think twice about it. Now I want to ask a welder some questions. (Turn to a welder) Do you work with both gas welding and electrical welding?

C_1: Yes, I do.

A: Then, if the gas hose in use has a flashback, what will you do to deal with it?

C_1: The hose should be replaced because it can sustain unseen internal damage.

A: If a back fire occurs in your torch, how to do with it?

C_1: I will close the torch valves, and if the torch is overheated, plunge it into water until it is cool.

A: OK, good. Can both electrical welding machines be connected to one switch board?

C_1: No, they can't. Each machine can only be connected to one switch board only.

A: Good job. Thank you.

(Coming to another worker.)

A: Excuse me. Have you ever worked in deep well or sewers?

C_2: Yes, I have. I worked in a sewer for a week last month.

A: Will you do any preparations before going down there?

C_2: Hardly. I will go down to work immediately when I am asked to.

A: Oh no, it is really dangerous to do so. It is important that you make sure there is enough fresh air in it before going down there. Please bear it in mind, and be careful next time, will you?

(Showing A around a working site where there is scaffolding.)

B: Here is our scaffolding. Since it was erected for external wall works, we have an inspection to all items of the scaffolding already, such as foundation, connection to wall, walk path and rail, bracing, ladders, loading test and so on.

A: All right, you've really done a good job, but I have to say some aspects have yet to be improved in the working condition above ground level. Firstly, some working tables which are over 2 meters from ground are not equipped with ladders. Secondly, in a few places there are too many materials piled on the scaffoldings.

B: All right. It is really a dangerous operation. We must rectify them as soon as possible.

A: Thirdly, in some opening of floors there are neither rails nor sufficient covers. All such things do not comply with the safety regulations and should be improved quickly.

B: OK, What you have mentioned have been recorded and they will be modified within a short time.

New Words and Expressions:

principle *n.*	原则,准则
to start with	首先
appoint *v.*	任命,委派
site management	工地管理
sufficient *a.*	充足的,丰富的
authority *n.*	权利,权威
ensure *v.*	确保,保证
in association with	与……相联系
rectify *v.*	纠正,调整
recommend *v.*	推荐,建议
full time	专职,专任
in charge of	主管,负责
additional bonus	额外奖金
foremen *n.*	工头,领班
supervisor *n.*	主管,监理
respiratory	呼吸的,与呼吸有关的
toxic substance	有毒物质
first aid and emergency aid procedure	急救和紧急援助程序
covers a wider range of	涵盖更广范围的
implement *v.*	实施,执行
negligence *n.*	疏忽,玩忽行为
in accordance with	依照,与……一致
helmet *n.*	头盔,防护帽
personal protective equipment	个人防护用品
discipline *n./v.*	管制,使守纪律

temporary a.	暂时的,临时的
electrical installation	电气装置
transformer n.	变压器
switch panel	配电板,开关屏
in a... manner	以……的形式
timing switch	定时开关,计时开关
short circuit	短路
electrical distribution	配电
fuse box	保险丝盒
copper wire	铜线
inadequacy a.	不恰当,不足之处
fire extinguishing equipment	消防设备
charged a.	盛满的,装满的
gas cylinder	高压气瓶,气缸
combustible a.	可燃的,燃烧性的
flammable a.	易燃的,速燃的
conspicuous a.	引人注意的,明显的
potential safety hazards	安全隐患
welder n.	焊接工
gas welding	气焊
electrical welding	电焊
gas hose	天然气软管
flashback n.	回火
torch n.	气炬,焊灯
plunge into	投入,浸入
switch board	电键板,配电盘
deep well	深井
sewer n.	下水管道
scaffolding	脚手架,施工架
bracing n.	拉条,紧固装置,支撑系统
loading test	载重试验;带负荷试验

Notes to the Text：

1. Is everybody responsible for safety from each section present? 各部门的负责人都到了吗?

2. Up to now, all documents regarding safety requirements and standards from the employer have been issued. 至今为止，业主已将所有有关安全要求和标准的文件都送交给了承包商。

3. Of course, penalty actions are the special cases we have to take. 当然,采取处罚也

是个别特例。

4. Our Safety Training Program is carried out on the basis of the Chinese Construction Ministry Standard JGJ59—88. 我们的安全培训计划是按照国家建设部标准 JGJ59—88 制定的。

5. I'm afraid this can not justify working without protecting feet. 我认为这不足以成为工作中不穿安全鞋的理由。

6. I believe the situation will be greatly improved with our joint efforts. 我相信在我们的共同努力下目前状况会得到显著改善。

7. In case any short circuit happens, the relevant switch will cut off electricity immediately. 一旦出现短路,相关的开关会立刻切断电源。

8. All criticisms are welcome and they are valuable to the improvement of our work. 我们希望得到您的批评指正,它对于我们改进工作具有重要的意义。

9. They are very likely to cause fire accident. 它们很可能会引起火灾事故。

10. In this way, smoking ban in the working areas can be effectively practiced. 这样的话,禁烟令也能得到更好地实施。

11. You'd better think twice about it. 你最好再认真考虑一下。

12. ... but I have to say some aspects have yet to be improved in the working condition above ground level. 我得说地面上的工程状况确实还有待提高。句中 have yet to do sth. 表示"还没有做某事"。例如:This theory seems to offer a better explanation to this phenomenon, but it has yet to be proved. 这个理论似乎能够更好地解释这一现象,但它仍有待证实。

Post-Reading Exercises:

Translate the following sentences into Chinese.

1. The first aspect is to give regular inspection of the site safety in association with the works supervision system, to ensure safe methods of work are being used, and all safety and health requirements are being observed.

2. The safety officer's second duty is to make detailed investigations to determine the cause of accidents and dangerous occurrences, and recommend possible solutions to prevent recurrences of similar accidents.

3. The training contents include many aspects such as jobsite safety policy, employee responsibility under China Labor Law, company's safety regulations, construction accidents and reporting, electrical safety, personal safety equipment, crane safety, respiratory protection, fire protection and prevention, toxic substances, first aid and emergency aid procedure, and so on.

4. Since it was erected for external wall works, we have an inspection to all items of the scaffolding already, such as foundation, connection to wall, walk path and rail, bracing, ladders, loading test and so on.

Part 5   Extensive Reading

(A) Functional Considerations in Design

Several functional principles are common to design regardless of the building type, efficient circulation of people, for example, is almost always a goal. "efficient" may be defined slightly differently in some cases. Usually it means that paces are planned so that the minimum possible distance is traveled by the users of the building. In a store it means that customers move in a pattern that encourages them to buy the most goods. Generally, however, walking between work stations or rooms in a home is considered wasted time and also tiresome.

Every square foot of hallway beyond the absolute minimum adds unnecessary cost to the construction, monthly utility bills and maintenance. Consequently, hallways and traffic paths through rooms should be kept to a minimum. The least useful hall is one with merely an opening at each end. A more useful hall is one with many additional doors and functions along the sides. An ideal use of a hall in a home is to have it double as a room such as a laundry. For example, if a washer and dryer are enclosed with a folding door on one side of a hall, the hall becomes a laundry room when the door is opened. A linen closet opposite the laundry alcove is handy to the washer and dryer and further increases the usefulness of the hall.

Hallway space requirements can be reduced by the way adjoining rooms are oriented. When placing a room at the end of a hall, orienting the short dimension of the room perpendicular to the hall will reduce the length of the hall. If the hall passes by a room, the length of the hall can be reduced by orienting the long dimension of the room perpendicular to the hall. When possible, plan the traffic pattern through a room so that it is perpendicular to the long dimension of the room. In this way, less of the room will be lost to traffic space.

Even if a hall can not have double use, it may be useful as a visual extension to a room. If the requirements for privacy will allow it, remove one wall of a hall so that the adjacent room seems larger. This may be possible more frequently in a home. In a commercial building, a glass wall may be used between a room and a hall to give both the room and the hall a feeling of being larger. The apparent length of an office hallway may be reduced by orienting fluorescent light fixtures across the width of the hall. If oriented lengthwise, the hall will appear longer.

Halls may serve as buffers between noisy and quiet spaces. By positioning a noisy room across a hall rather than adjacent to a quiet room, sound transmission to the quiet room will be reduced.

A short straight hall will keep the number of turns a person must negotiate to a minimum. If a hall must turn, keep the configuration simple. Irregularly shaped halls are inconvenient to use. They may also be a sign of inefficiently planned rooms and unnecessarily complicated construction.

Another goal typical of both commercial and residential design is to separate private and public activities. In residential planning, the distinction between the two is subtle since a home is a private building. Living and dining areas are often used to receive visitors. The bedrooms, however, are not as public. This affects the placement of rooms so that non-residents of the home aren't forced to pass through the private area in the course of their visit. It may be desirable to provide a toilet for guests separate from the bathrooms serving the bedrooms. Both private and public areas may be in use at the same time, making their separation an important practical matter.

In businesses where there are areas for employees only, the separation between public and private use is easily accomplished. Walls and locked doors are clear barriers. In situations where limited public access to private areas is allowed, the design must provide for control of the access but not make it difficult. A receptionist can provide such control if the rooms are arranged properly. The presence of a receptionist or secretary should be an implied physical barrier, so that it is apparent that permission is required to pass.

The need for acoustical privacy may be met in a home with the same plan that provides physical privacy for the bedrooms. If the bedrooms are grouped together and separated from public areas, noise is less likely to be bothersome to those sleeping while activity continues elsewhere in the house. Acoustical privacy in commercial buildings is not usually solved by the private/public separation.

The need for solving noise problems in nonresidential buildings ranges from little to very important. Some businesses do not generate noise, or at least the noise level in all spaces is about equal. Offices with few typewriters and no machines may simply need privacy for conversation. Several business machines in use will require at least a separate closed room. The location of doors is also an important consideration in controlling noise.

Light manufacturing may need to be located at one end of a building while the office functions take place at the other. Heavy manufacturing may require a building separate from the offices.

Increasing parts of a building's cost are the service components. If you can achieve other design goals while increasing the efficiency of the heating, air conditioning, electrical and plumbing systems, it is obvious that you should do so. Since plumbing involves only certain rooms, the position of these rooms is important to reduce the cost of the plumbing system. Kitchens (or kitchenettes), toilets and water heaters are common in many buildings. It is standard practice to try to bring these items together to reduce the cost of piping and the labor to install it. Features such as drinking fountains, janitor's closets and other commercial plumbing needs are, of course, included in this grouping when they are present. In multistory buildings, the stacking of rooms having plumbing is a standard procedure. The savings in a single small house may not be great, but their small savings multiplied by a hundred units in a subdivision or in an apartment building becomes significant.

Notes to the Text:

1. efficient circulation of people: 效率高的人行路线。
2. to have it double as a room: 使它具有双重功能（既是厅又是房间）。
3. the laundry alcove: 洗衣间。
4. by the way adjoining rooms are oriented: 根据相邻房间的朝向去设计。
5. as a visual extension to a room: 作为房间的一种视觉上的延伸。
6. by orienting fluorescent light fixtures across the width of the hall: 通过日光灯装置横向悬挂在厅的上面。
7. if oriented lengthwise: 如果纵向设置。
8. A short straight hall will keep the number of turns a person must negotiate to a minimum: 短的直通厅将使人必须通过的转弯口降至最低限度（negotiate 表示"通过"；keep... to a minimum: 把……降至最低限度）。
9. an implied physical barrier: 一道暗示性的身体屏障（表示禁止通过）。
10. The need for solving noise problems in nonresidential buildings ranges from little to very important: 解决非居住用房的噪音问题，其迫切性相差甚大，从基本无此要求，到有迫切的需要。
11. the stacking of rooms having plumbing is standard procedure: 把有管道的房间上下重叠起来是人们普遍采用的设计程序。

New Words and Expressions:

functional	a.	功能的，实用的
principle	n.	原理
circulation	n.	循环
minimum	n.	最小量，最低限
hallway	n.	门厅，走廊
utility	n.	有用，实用
maintenance	n.	保持，维持
dimension	n.	体积，程度
perpendicular	a.	成直角的，垂直的
fluorescent	a.	荧光的，发亮的
buffer	n.	缓冲器，减震器
adjacent	a.	与某物邻近的，接近的
subdivision	n.	进一步细分

configuration	n.	某物的构造,结构
commercial	a.	商业的
residential	a.	适于住宅的
subtle	a.	难以描述的,细微的
desirable	a.	称心的
accomplish	v.	完成,实现
acoustical	a.	声音的,音响的
privacy	n.	独处,隐私
generate	v.	产生
component	n.	组成部分,成分,零部件
plumbing	n.	测深锤,铅锤
item	n.	项目,条款
install	v.	安装,安置
closet	n.	柜橱,小房间

Posts-Reading Exercises:

1. Choose the one that best completes the sentence.

1) In a kitchen, the traffic between work _____ should be kept to a minimum.
   A. sites   B. fields   C. cell   D. den

2) A(n) _____ is a small partly enclosed space in a room for a bed, books and a washer, etc.
   A. alcove   B. corner   C. cabinet   D. container

3) A house can be designed so that a visual _____ to a room will be achieved.
   A. expansion   B. extension   C. expression   D. expectation

4) Halls are often used as _____ between various spaces.
   A. barriers   B. buffers   C. entries   D. openings

5) The distinction between private and public activities is not as _____ in nonresidential planning as in residential planning.
   A. detailed   B. confused   C. subtle   D. direct

6) The toilet should be positioned so that public _____ to private areas is not allowed.
   A. admission   B. arrival   C. access   D. appearance

7) Rooms should be planned so that noise _____ from one place will not be easily transmitted to another.
   A. generated   B. formed   C. discovered   D. gained

8) Noise is likely to be _____ to those sleeping while construction is going on outside the home.
   A. extreme   B. effective   C. bothersome   D. annoyed

9) _____ is one of the elements which should be considered in planning a building.

165

A. Privacy B. Tradition C. Control D. Dimension

10) The cost of a building is associated with the hallways and traffic paths through rooms which should be kept _____ .

A. lowly B. efficiently C. to a degree D. to a minimum

2. Answer the following questions.

1) What does efficient circulation of people usually mean?

2) What will the unnecessary hallway space in a building result in?

3) Which is more useful, a hall with merely an opening at each end or a hall with many additional doors along the side?

4) How can the length of a hall be reduced when a room is placed at the end of the hall?

5) To make a room appear larger, what can we do with the adjacent hall?

6) In regard to business, what can be done to separate private areas from public ones?

7) How can acoustical privacy be solved in planning a home?

8) How can we reduce the cost of service components in a building?

(B) Making Architectural Judgments

How should an architect, a client, a citizen, or a government agency or commission judge designs for new buildings? This question is especially pertinent if the new structures are to be erected in historic districts. How does one decide whether the proposed new building will enhance or detract from its surroundings, and whether it promotes the kind of further development that will benefit the historic area? What criteria can be used as a basis for such judgments? Do the answers lie in adhering to some, "correct" architectural ideology derived from a classical, a modernist or a post-modernist point of view to be handed down by "experts"? Does the task require studied connoisseurship, group consensus growing out of broad public participation, or some combination of these positions? (A)

The need for a more rational approach to these questions is evident from the nature of the discussion that has surrounded recent controversial buildings, in which reasonable differences of opinion have often devolved into vituperative confrontations. The differences among opposing groups often hinge on passionately held beliefs about aesthetics, politics or such vague notions as the demands of the Zeitgeist, the "spirit of the times". The focus on subjective questions of taste, ideology and personality tends to discourage constructive debate and to ignore more complex and pertinent questions of how to protect the public interest. (B)

The depth of this problem may be seen by reviewing the weak and conflicting statements of problems that are presented within transcripts of hearings before boards and commissions with purview over architectural projects, and by reading their subsequent reports. Anyone who has attended juries evaluating students' work at schools of architecture will be familiar with the subjective criteria that often pass for considered, objective judgments. Because decision-making bodies cannot evade their responsibilities, this haphazard, emotion-laden way of defining architectural standards has created the incoherently planned cityscape, the suburban sprawl, and the suburbanized countryside we see around us. (C)

This contemporary dilemma should be considered in its proper historical context. The generations of architects since 1945 are the first in the history of architecture who have not been able to fulfill two expectations that society has taken for granted since the ancient Greeks created cities: that architects are able to create both the competently designed background architecture that forms the bulk of building in a town or city as well as noble foreground buildings, the monumental architecture of civic and religious structures that embodies a society's highest aspirations. It was in this way that architects of the past arranged their designs in city plans so as to make the new as beautiful or better than what was there before. (D)

Architecture is a liberal art that is taught at our great universities, and is a discipline that should be amenable to rational discourse. The aim here is not to decide on the ultimate artistic worth of a proposed building, for this requires the kind of considered assessment that, ultimately only the distance of time can provide. Instead, it is to slightest a process that may assist us in evaluating the quality of proposed designs for new architecture in a variety of historic areas, a way of deciding whether the proposal at hand will add to or detract from the beauty and character of the place. The process offers a set of criteria, which can help to make distinctions between the architectural characteristics of a proposed building and to relate them to those of the buildings, which exist around its site. Once articulated, these distinctions provide a common basis for concerned citizens, critics, decision-making bodies and architects to debate the drawbacks and virtues of the design and to prepare a solid foundation for a decision to accept or to improve or otherwise modify the design or program. (E)

Architecture is a public art. It is the building block of the city, a compound work, realized over centuries. Cities are always changing, growing and being altered, being destroyed and being rebuilt, all in response to social and political change, to the demands of commerce and industry, and to the rhythms of technological innovation. Villages and rural areas are subject to the same pressures. Each time a new building is proposed it must be studied, not in isolation, but evaluated as a part of the increasingly complex whole that forms all regions. (F)

Notes to the Text:

1. "the depth of this problem may be seen by... and by...": 由 by 引出的介词短语作方式状语;"with purview over"在这里意思是"有……权限"。

2. decision-making bodies: 决策机构(关)。

3. as beautiful, or better than, what was there before: 在 beautiful 后省去了 as, 在 as... as 结构中第二个 as 引出的比较状语从句可省略部分或全部, 包括 as 本身。

New Words and Expressions:

detract	v.	减损; 降低
adhere	v.	坚持; 黏附(常跟 to)
ideology	n.	思想体系; 意识形态
connoisseurship	n.	鉴赏能力
consensus	n.	一致
devolve	v.	转移, 退化
vituperative	a.	责骂的
confrontation	n.	对抗, 对立
hinge	v.	依……为根据
zeitgeist	n.	(德) 时代精神
transcript	n.	副本; 文字记录
hearing	n.	意见听取会
purview	n.	权限
jury	n.	评审员; 陪审团
evade	v.	逃避, 避开
haphazard	a.	任意的, 杂乱的
subjective	a.	主观的
emotion-laden	a.	带情绪的
incoherently	ad.	前后矛盾的
sprawl	n. v.	(无规则地)蔓延
suburbanize	v.	市郊化
dilemma	n.	困境, 进退两难
civic	a.	城市的; 市民的
campanile	n.	钟楼
abound	v.	大量存在
fulcrum	n.	支点
propylene	n.	(神殿等)入口
revitalize	v.	使新生, 使恢复元气
amenable	a.	经得起检验的
discourse	n. v.	论述; 谈论
articulate	v.	表现(思想); 明确表达
drawback	n.	欠缺; 障碍

Post-Reading Exercises:

Match each of the following ideas with its appropriate paragraph number A—F.

1) People hold different opinions about how to evaluate designs for new buildings.

( )

2) Architecture in history is still of great significance today. ( )

3) Subjective criteria are often adopted for objective judgments. ( )

4) Architectural judgments should be made in accordance with the overall development. ( )

5) The paragraph that tells us the purpose of the author in writing this essay.

( )

UNIT 11   MEASUREMENTS

Part 1   Warm-up Activities

1. Look and read:

A room has three spatial dimensions: length, height and width. These dimensions are measured in millimeters or meters. The volume of a room equals length times height times width. Volume is measured in cubic meters(m^3). The area of a surface in the room is measured in square meters(m^2).

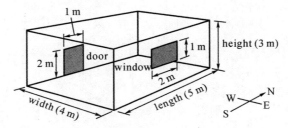

Internal measurements of a room

Make questions and answers about the volume, longitudinal-sectional area, cross-sectional area and surface areas of the room:

Example: *What is the internal area of the north-facing wall?*

The internal area of the north-facing wall is 12 square meters.

Now make statements like this:

The north-facing wall has an internal area of 12 square meters.

2. Estimate the internal measurements of your classroom and make a table like this:

Object	Dimension	Quantity	Unit
classroom	width	4	meters
window	area	2	square meters

3. Describe your classroom. Use the table of exercise 2:

The classroom has a width of approximately 4 meters.

The classroom is approximately 4 meters wide.

The window has an area of approximately 2 square meters.

The window is approximately 2 square meters in area.

Part 2 Controlled Practices

1. Look and read:

The architect has to consider the maximum, minimum and average dimensions of the human body to design buildings in the right scale. Decide for each design situation shown below, which dimension an architect should base his calculations on:

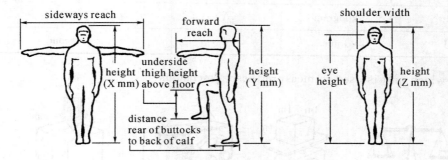

In a group of three people, their height $\begin{Bmatrix} \text{varies between} \\ \text{ranges from} \end{Bmatrix}$ X $\begin{Bmatrix} \text{and} \\ \text{to} \end{Bmatrix}$ Z mm. The maximum height is Z mm and the minimum is X mm. The average height in this group is therefore $\left\{\dfrac{X+Y+Z}{3}\right\} = W$ mm.

Now make a table of the following measurements of each person in your group and then write three similar paragraphs:

	Height	Eye height	Forward reach	Shoulder width	Length of lower leg	Length of upper leg
	in mm	in mm	in mm	in mm	in mm	in mm
Student 1						
Student 2						
Student 3						
etc.						

2. Use the measurements you collected in exercise 3 to help you complete this paragraph:

When deciding on the floor-to-ceiling height of a building, an architect should base his calculations on the tallest person in a group of people. The tallest person in our group is _____ mm. Therefore the floor-to-ceiling height of our building should be greater than _____ mm.

3. Common measurements in architecture:

Unit symbol	SI unit	Unit symbol	SI unit
lm	lumen	kg/m³	kilogram per cubic meter
lx	lax(1 lumen/m²)	N/mm²	Newton per square millimeter
°C	degree Celsius	dB	decibel
kg	kilogram	A	ampere
J	Joule	s	second

Say which unit is used to measure these dimensions:

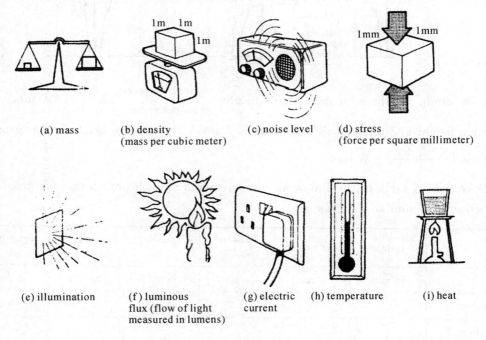

(a) mass
(b) density (mass per cubic meter)
(c) noise level
(d) stress (force per square millimeter)
(e) illumination
(f) luminous flux (flow of light measured in lumens)
(g) electric current
(h) temperature
(i) heat

Part 3 Further Development

1. Look at this diagram and complete the sentences below:

a) The temperature of the room is measured in _____ .
b) The _____ in the concrete block is 5 N/mm².
c) The noise level of the television is measured in _____ .
d) The illumination in the room is measured in _____ .
e) The _____ of the concrete foundation is 2,400 kg/m³.
f) The electric current to the television is measured in _____ .
g) The _____ from the light bulb is 50 lux.
h) The heat flow rate of the radiator is measured in _____ .

2. Now make a list of some of the things in your classroom that can be measured. Say what they are and what units they are measured in.

3. Now read these problems and complete the solutions:

a) *Problem*

A family of 5 persons wants to build a house. The floor area allowed for each person is 10 square meters. What is the floor area required?

Solution

There are _____ people. The floor area allowed per person is _____ _____ _____ .

Therefore _____.

b) *Problem*

An architect wants to build a concrete column to take a compressive force of 2,000 Newton. The maximum compressive stress allowed in the concrete is 5 N/mm². What is the minimum cross-sectional area of the column required?

Solution

The load on the column is _____ _____ . The _____ _____ allowed in the concrete is _____ _____ . Therefore _____.

c) *Problem*

An architect wants to build a concrete wall with a volume of 10 cubic meters. The maximum weight of the wall allowed is 22,000 kilogrammes. What is the maximum density of concrete required?

Solution

The volume _____

Therefore _____.

Part 4　Business Activities

1. 背景与指南(Background and Directions)

近年来,随着社会的发展,环境保护正日益成为全球关心的问题(global concern)。各国政府对在本国工程项目建设中可能造成的污染都很重视,常常在工程施工过程中派有关人员到工地去检查。项目的业主也常在合同中对施工中的环保问题作出相应的规定。

承包商根据业主的要求往往也编制自己的环保计划(environmental protection plan),对开挖出的弃土与石渣的倾倒(dumping of surplus soil and muck)、施工现场上污水(sewage)与垃圾(garbage)的处理,特别是对于一些可能出现化学污染(chemical pollution)防范措施等作出详细的规定,供业主批准。

2. 情景会话(Situational Conversation)

Environmental Protection

(*As environment has become a big concern in the modern world, environmental protection is always emphasized by the country where the works are executed. At today's meeting, matters on the project's environmental protection are scheduled to be discussed.*)

(*A: The owner's representative, B: the contractor's project manager, C: Inspector from the local environmental protection agency.*)

A: Mr. Lin, I don't think you've met Mr. Lucas, inspector of EPA. Mr. Lucas. This is Mr. Lin, project manager of the Contractor for this project.

B: Welcome to the project, Mr. Lucas. It's an honor to meet you.

A: Mr. Lucas hopes to learn how the environment is protected while the project's going on.

C: Yes, that's what I am here for. This is a very important project for our country. However, some local residents from the nearby community are rather concerned with its possible adverse impact on the environment during the construction. What environmental protection measures do you undertake?

B: Mr. Madison and I have discussed this matter several times. Both sides attach great importance to it. Right from the beginning of the works, we studied the environmental protection regulations published by the Government. Accordingly, we have made a careful plan on this project, where guidelines are set out for all the employees on this project to observe for protecting the relevant land, surrounding vegetation and, especially, the river water.

C: May I have more specific details of your measures?

B: To avoid contamination by the leaking of fuel and garbage, various treatment facilities have been built. Latrines, which lose to the working areas, were erected in accordance with Mr. Madison's suggestions. Protection of the river from contamination is where we spend our more effort. Things like empty cement bags, cartons and other waste materials are cleared away from the river banks in time to prevent them from falling into the river. As to the waste water from the crushing system at our batch plant, special treatment is carried before it is let go into the river.

C: Could you elaborate on your special treatment?

B: We have a settlement tank system. it is cellular with three cells. The waste water from the sand and aggregate washer first runs along the outlet ditch with filtering devices into the first cell for initial settlement. After an interval of twenty four hours, the water flows into the second cell where chemical treatment is given for the rock flour to settle. Finally, the treated water reaches the river through the third cell. The sediment sludge is cleaned out regularly and disposed of in a proper area.

C: Sounds fine, but how do you guarantee that this plan is implemented?

B: We have designated one of our staff to be solely in charge of this job. He is responsible for monitoring and examining all the working crews carrying out this work. Of course, we are always under the superintendence of the owner.

A: That's right.

C: Can I go to the job site and have a look, please?

A: You are welcome to visit the job site at any time you like.

B: After your visit to the site, you could go back and tell those worried residents that this project will not do any harm to their lives. Instead, it will improve their living conditions.

New Words and Expressions:

adverse impact	不利影响
contamination *n.*	污染
leaking *a.*	漏,渗
treatment facility	(废物)处理设施
carton *n.*	纸板箱

crushing system	碎石系统
elaborate　v.	详细说明
settlement tank system	沉淀池系统,沉沙池系统
cellular　a.	网格状的
cell　n.	单元格,区格
sand and aggregate washer	沙子与骨料冲洗机
outlet ditch	排水沟
filtering device	过滤装置
rock flour	石粉末
settle　v.	沉淀
sediment sludge	沉积泥沙
dispose of　v.	处理
designate	指定
under the superintendence of	在……的监督下

Notes to the Text:

1. it's an honor to meet you：见到你真荣幸。这是首次见面时常说的客套话。

2. Mr. Lucas hopes to learn how the environment is protected...：Lucas 先生希望了解一下环保护方面的问题……句中的 learn 在这里是"了解"，"得知"的意思。例如：When he learned that the goods could not be delivered on time, he made a call to the supplier immediately himself.（当他得知货物不能按时交付他,立即亲自给厂家挂了电话。）

3. EPA：环境保护局 Environmental Protection Agency 的缩写形式。

4. What environmental protection measures do you undertake? 你们采取的环保措施有哪些？undertake 在这里等于 take。

5. nearby community：附近的居民 community 意思是"团体"，"社区"，"同一地区的全体居民。如：international community（国际社会），religious community（宗教团体）。

6. Right from the beginning of the works：从工程一开始。right 这里是加强语气的副词。

7. garbage：废料,垃圾。英国人常用 rubbish 一词。

8. latrine：工地上的简易厕所

9. initial：初始的,第一次；在合同管理中,这个词用作动词时,意思是"草签"、"小签"，指在文件上签上自己名字的首写字母,如：名字为 John Brown,则签 JB,中国人一般签自己的姓（last name）。

10. ...this project will do not any harm to their lives.：这一工程项目不会对他们的生活产生危害。do harm to...意思是对"对……有害"；do no harm to...意思是"对……无害"。例如：Poor ventilation will do harm to the workers' health.（通风不好会损害工人的健康。）

Post-Reading Exercises:

Do the following translation:

1. "Could you elaborate on your special treatment?"(你是否再详细地介绍一下你们专门的污水处理方法?)

"Could you elaborate on...?" is a very polite way to ask for more detailed information about something, meaning "您是否能详细地说明一下……"

1) A: We'll do something about it.
   B: _____.
   (您是否能详细说一下你们的具体措施?)

2) A: This is our tentative plan.
   B: _____.
   (我们希望你们编制得再详细点。)

3) A: _____.
   (你们只需告诉我们事实,不必加以解释。)
   B: All right, but the facts must be interpreted impartially.

2. "You are welcome to visit the job site at any time you like.(随时欢迎您到工地参观。)"You are welcome to do sth." is an expression to welcome someone to do something, meaning "欢迎您……" or "您尽管……".

1) _____
   (欢迎您到我们公司参观。)

2) _____
   (欢迎您提出自己的建议。)

3) _____
   (这台设备您尽管用。)

Part 5 Extensive Reading

(A) Concrete Strength

Concrete is made from cement, coarse aggregate (stones), fine aggregate (sand or crushed stone) and water. Coarse aggregate ranging from 5mm to 40mm may be used for normal work. The maximum size of the aggregate should not be greater than one quarter of the minimum thickness of the finished concrete. The normal maximum sizes are 20mm and 40mm, 20mm being more common. The maximum size of aggregate which should be used in small concrete sections, or where reinforcement is done together, is 10mm.

In concrete with widely spaced reinforcement, such as solid slabs, the size of the coarse aggregate should not be greater than the minimum cover to the reinforcement

Reinforced concrete section

otherwise spilling will occur, i. e. the breaking off of pieces of concrete below the reinforcement. For heavily reinforced sections, e. g. the ribs of main beams, the maximum size of the coarse aggregate should be either(take the smaller one): (1) 5mm less than the minimum horizontal distance between the reinforcing rods, or(2) 5mm less than the minimum cover to the reinforcement.

Concrete is a very strong material when it is placed in compression. It is, however, extremely weak in tension. It is for this reason that we use reinforcement in concrete structures. The reinforcement, which is usually steel, takes up the slack for the weakness of the concrete in tension.

There are many ways to test the strength of a batch of concrete. The tests used can be categorized as destructive and nondestructive tests. Usually when a batch of concrete is ordered on a job site it is specified to be of a specific compressive strength—4,000 psi, for instance. When the concrete comes to the job site in a ready-mix truck, the contractor places some of the batch in cylinders which are 6 inches in diameter and 12 inches in height. These cylinders are cured for 28 days and tested by compression until they are crushed. This will give the contractor or the engineer the compressive strength for that batch of concrete. He or she can then compare that value to the design value used to make sure that the structure was constructed properly.

Once the concrete has been placed for a particular structure, there is a nondestructive test which can be performed to estimate the strength of the concrete. This method uses a Schmidt hammer(also called a Swiss hammer). This method of testing is based on the inertia of a ball inside the Schmidt hammer testing apparatus that is "bounced off" of the concrete.

Post-Reading Exercises:

1. Now decide whether these statements are true(T) or false(F). Correct the false statements.

　　1) Concrete is made from three different materials.　　　　　　　　　　　(　　)

　　2) Coarse aggregate ranges in size from 20mm to 40mm.　　　　　　　　(　　)

　　3) When the minimum thickness of the finished concrete is 100mm, the maximum size of aggregate should not be greater than 25mm.　　　　　　　　　　　　　(　　)

　　4) When the reinforcing rods are close together, the maximum size of aggregate used should be 10mm.　　　　　　　　　　　　　　　　　　　　　　　　　(　　)

　　5) Cover is the thickness of concrete between the reinforcing rods.　　　　(　　)

6) Spilling can occur in a solid concrete slab when the cover to the reinforcement is greater than the maximum size of the coarse aggregate.   (   )

7) When the minimum horizontal distance between reinforcing rods is 15mm, the maximum size of aggregate should be less than 12mm.   (   )

2. Answer the following questions:

1) Is concrete a very strong material? And in what condition is it extremely weak in tension?

2) What method do we use to reinforce the concrete structures?

3) How many ways are there to test the strength of a batch of concrete?

4) How many days are the cylinders cured?

5) What kind of tests are the cylinders endured?

6) Why does the contractor or the engineer want to know the compressive strength for that batch of concrete?

(B) Surveying

Before any civil engineering project can be designed, a survey of the site must be made. There are two kinds of surveying: plane and geodetic. Plane surveying is the measurement of the earth's surface as though it were a plane (or flat) surface without curvature. For larger areas, a geodetic survey, which takes into account the curvature of the earth, must be made.

The different kinds of measurements in a survey include distances, elevations (heights of features within the area), boundaries (both man-made and natural) and other physical characteristics of the site. Some of these measurements will be in a horizontal plane; that is, perpendicular to the force of gravity. Others will be in a vertical plane, in line with the direction of gravity. The measurement of angles in either the horizontal or vertical plane is an important aspect of surveying in order to determine precise boundaries or precise elevations.

In plane surveying, principal measuring device for distance is the steel tape. In English-speaking countries, it has replaced a rule called a chain, which was either 66 or 100 feet long. The 66-foot-long chain gave speakers of English the acre, measuring ten square chains or 43,560 square feet as a measure of land area. The men, who hold the steel tape during a survey, are still usually called chainmen. They generally level the tape by means of plumb bobs, which are lead weights attached to a line that give the direction of gravity. When especially accurate results are required, other means of support, such as a tripod can be used. The indicated length of a steel tape is in fact exactly accurate only at a temperature of 200centigrade, so temperature readings are often taken during a survey to correct distances by allowing for expansion or contraction of the tape.

Distances between elevations are measured in a horizontal plane. In the diagram along side, the distance between the two hills is measured from points A to B rather than from points A to C to D to B. When distances are being measured on a slope, a procedure called breaking chain is followed. This means that measurements are taken with less than the full length of the tape. Lining up the tape in a straight line of sight is the responsibility of the transit-man, who is equipped with a telescopic instrument called a transit. The transit has plates that can indicate both vertical and horizontal angles, as well as leveling devices that keep it in a horizontal plane. Cross hairs within the telescope permit the transit-man to line up the ends of the tape when he has them in focus.

Angles are measured in degrees of arc. Two different systems are in use. One is the hexadecimal system that employs 3,600, each degree consisting of 60 minutes and each minute of 60 seconds. The other is the centesimal system that employs 400 grads, each grad consisting of 100 minutes and each minute of 100 seconds. A special telescopic instrument that gives more accurate readings of angles than the transit is called a theodolite. In addition to cross hairs, transits and theodolites have markings called stadia hairs. The stadia hairs are parallel to the horizontal cross hairs. The transit-man sights a rod, which is a rule with spaces marked at regular intervals. The stadia hairs are fixed to represent a distance that is usually a hundred times each of the marks on the rod. That is, when the stadia hairs are in line with a make on the rod that reads 2.5, the transit is 250 meters from the rod. Stadia surveys are particularly useful in determining contour lines, the lines on a map that enclose areas of equal elevation.

Contour maps can be made in the field by means of a plane-table alidade. The alidade is a telescope with a vertical circle and stadia hairs. It is mounted on a straight-edged metal plate that can be kept parallel to the line of sight. The surveyor can make his readings of distances and elevations on a plane(or flat table that serves as a drawing board). When the marks representing equal elevations are connected, the surveyor has made a contour map. Heights or elevations are determined by means of a surveyor's level, another kind of telescope with bubble leveling device parallel to the telescope. A bubble level is a tube containing a fluid that has an air bubble in it. When the bubble is centered in the middle of the tube, the device is level.

Geodetic surveying is much more complex than plane surveying. It involves measuring a network of triangles that are based on point on the earth's surface. The triangulation is then reconciled by mathematical calculations with the shape of the earth. This shape, incidentally, is not a perfect sphere but an imaginary surface, slightly flattened at the poles, that represents mean sea level as though it were continued even under the continental landmasses.

New Words and Exercises:

civil engineering	土木工程
survey *v. n.*	测量;测量学
route *n.*	路线

plane	*a.*	平面的
geodetic	*a.*	大地测量（学）的
curvature	*n.*	弯曲,曲度,曲率
elevation	*n.*	高度,高程,海拔
boundary	*n.*	边界,分界
perpendicular to		与……垂直,成正交
steel tape	*n.*	钢尺
chain	*n.*	链条
acre	*n.*	英亩
plumb bob	*n.*	铅球
tripod	*n.*	三脚架
centigrade	*n.*	百分度,摄氏度
expansion	*n.*	膨胀
contraction	*n.*	收缩
transit	*n.*	经纬仪
cross hairs		十字丝
hexadecimal	*a.*	六十的,六十进位的
centesimal	*a.*	百分的
theodolite	*n.*	光学经纬仪
stadia hairs		视距丝
rod	*n.*	标尺,水平尺
contour line		等高线
alidade	*n.*	照准仪
parallel to		平行于
level	*n.*	水平面
bubble level		气泡水平仪
visibility	*n.*	可见的
bench mark	*n.*	水准基点
altimeter	*n.*	测高仪,高度仪
atmospheric pressure		大气压力
humidity	*n.*	湿度
target	*n.*	目标

Post-Reading Exercises:
1. Choose the one that best completes the sentence.
1) Plane surveying is means _____.
A. measuring a horizontal surface
B. measuring the earth's surface with curvature
C. determining the best and most economical location or rout

D. measuring the earth's surface without considering its curvature

2) When the areas are larger than 20 kilometers square, _____.

A. the earth's curvature does not effect the accuracy

B. the earth's curvature should be considered in surveying

C. a geodetic survey is unnecessary

D. a geodetic survey must take into account the curvature of the earth

3) When measuring the distances and elevations, the measurements will be in _____.

A. horizontal plane

B. vertical plane

C. either the horizontal or vertical plane

D. both the horizontal and vertical plane

4) The acre is _____.

A. the 66-foot-long chain

B. the area of measuring ten times of chains

C. the area of measuring 10 square chains

D. a measure of land area

5) When the temperature is higher than 20 centigrade, the indicated length of a steel tape is _____.

A. exactly accurate   B. longer than actual length

C. shorter than actual length   D. need to correct distance

6) The function of the cross hairs within the telescope is _____.

A. sighting the target   B. measuring the angles

C. measuring the distances   D. focusing on tape

7) The heights or elevations are determined by means of _____.

A. transit   B. altimeter   C. stadia hairs   D. level

8) Which is more complete?

A. Plane surveying   B. Geodetic surveying

C. Spot surveying   D. Distance surveying

2. Fill in the blanks with the proper phrases given below changing the form if necessary:

| as though   take into account   perpendicular to   equip... with... |
| as well as   by means of   parallel to   based on |

1. You will get involved, and do something to contribute to your school _____ to help yourself.

2. The reports will now _____ the international consequences of each nation's actions.

3. The pilot homed _____ in radar, despite the thick fog.

4. It was only three years ago, but it feels _____ centuries have passed.

5. Your torso(躯干) should now be _____ the wall, with the blankets under you.

6. They _____ high definition technology, digital tuners and touch screen functionality.

7. The writer's controversial Broadway play was, in fact, _____ a true story.

8. My experience in this matter is _____ yours, so I have nothing else to say.

UNIT 12   THERMAL AND MOISTURE PROTECTION

Part 1   Warm-up Activities

1. Look and read:

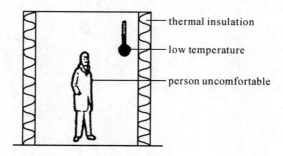

Question 1: Why is this person uncomfortable?
   Because the temperature in the room is too low.
   Because the room is excessively cold.

Question 2: Why is the temperature *too low*?
   Because the thermal insulation is *inadequate*.
   Because it has an *insufficient amount of* thermal insulation.
   Because the thermal insulation is *not* thick *enough*.

2. Look at these drawings and make similar pairs of questions and answers:

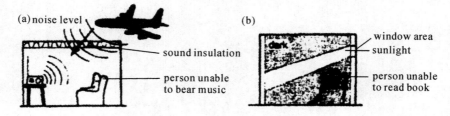

UNIT 12　THERMAL AND MOISTURE PROTECTION

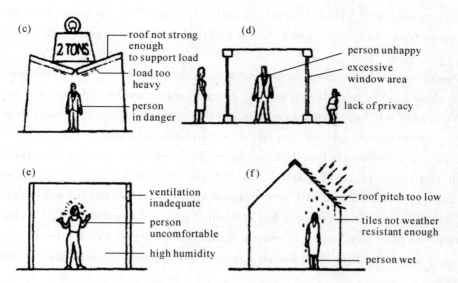

3. Make similar statements about thermal protection and sound control, using the phrases given below:

a) (designs size of windows) adequate window area/sufficient privacy
b) (designs window area) sufficient warmth/adequate light
c) (designs thickness of sound insulation) cheap enough materials/adequate sound insulation
d) (designs ventilation system) sufficient warmth/not excessive humidity
e) (designs roof) cheap enough materials/adequate weather proofing

Example:

When an architect <u>designs a house</u>, he often has to strike a balance between two conflicting requirements. For example, he needs to ensure that there is <u>adequate ventilation</u> and at the same time he needs to ensure that the <u>noise level</u> is not excessive.

Part 2　Controlled Practices

1. Read the passage, and fill in each of the following blanks with an appropriate word from the box.

Design of Houses for Tropical Climates

extending	excessive	comfort	shade	screening
occur	radiation	adequate	interior	consequently

The tropical regions of the earth can be divided into three major climatic zones:

A. *Warm-humid climates* are found in a belt near the Equator 1) _____ to about 15° north and south. There is very little seasonal variation throughout the year. The air temperature is never 2) _____, but there is considerable rainfall during most of the year. Relative humidity(RH) is excessively high—at about 75% for most of the time, but it may vary from 55% to almost 100% (RH should not exceed 70% for human 3) _____).

In this climate, the rooms of houses must have adequate 4) _____ and ventilation. Usually houses have an open layout so they can gain maximum benefit from the prevailing wind. Walls have less importance here than in other climates. They are used primarily for 5) _____ from insects and for their wind penetration qualities.

B. *Hot-dry climates* are found in(two belts of latitude between approximately 15° and 30° north and south of the Equator. Two marked seasons 6) _____ : a hot and a slightly cooler period. Daytime air temperatures are excessively hot(normally higher than the 31°C to 34°C skin temperature), but at night it may fall as much as 35°C. During the day there is too little cloud cover to reduce the high intensity of direct solar 7) _____ . However, at night the clear skies permit a considerable amount of heat to be reradiated to outer space.

In this climate houses must give 8) _____ protection against the excessive heat of the sun. Usually they have compact layouts, so that surfaces exposed to the sun are reduced as much as possible. Walls should be very thick and made of heat storing materials so they hold the heat of the day and give it back to the 9) _____ of the house at night.

C. *Composite or monsoon climates are* found in large land masses near the tropics of Cancer and Capricorn. Two seasons occur normally. Approximately two-thirds of the year is hot-dry and the other third is warm-humid. 10) _____ , houses designed to be suitable for one season may be unsuitable for the other.

2. Look at these diagrams of 2 different house types and decide which house type is most suitable for a hot-dry climate and which is most suitable for a warm-humid climate. Explain why, using information from the reading passage and the diagrams above.

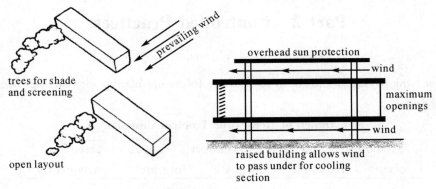

Type A

UNIT 12 THERMAL AND MOISTURE PROTECTION

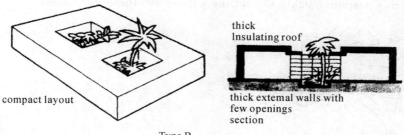

compact layout

thick Insulating roof

thick external walls with few openings
section

Type B

3. Now complete this passage by using these words given:

| hottest | coolest | cool enough | warm enough |
| excessively | much lighter | inadequately | adequately |

In composite climates, houses designed to perform _____ for one season will perform _____ for the other. To solve this problem, houses are sometimes built two storeys high. The ground floor is built with _____ thick walls. These retain the heat so that it is _____ to sleep comfortably on the ground floor during the _____ part of the year. The first floor structure is built with materials. This structure cools quickly at night so that it is to sleep comfortably on the first floor during the _____ part of the year.

Part 3 Further Development

1. Look at the table and use them to make conversations between the client and the architect, following the example below:

Example:

client: What do you think is the best material to use for the cladding?

architect: Well, aluminum isn't really suitable. It's strong enough, but too expensive. I think we should use mild steel. It's not only strong enough, but also cheaper.

Building component	Possible materials	Performance requirement
cladding	aluminum/mild steel	a tensile strength of not less than 90 N/mm^2
beam	pine/oak	a tensile strength of not less than 35 N/mm^2
roof covering	zinc/copper	weight should not exceed 50 kg/m^2
fire door	mild steel/copper	melting point should be in excess of 1,000℃

2. Read this passage quickly to find the answers to these questions:

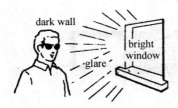

What causes glare in

a) hot-dry regions?

b) warm-humid regions?

One of the problems in hot climates is to exclude not only radiant heat but also glare, while at the same time admitting sufficient daylight. There is a fundamental difference between the problem in the arid and humid regions. In the arid regions, glare is caused by sunlight being reflected from the surface of the ground and light colored walls of other buildings. A traditional way of overcoming this problem is by keeping windows on the external elevations small and few in number, with the larger, low level windows overlooking the shaded internal courtyard. Too sharp a contrast between a bright opening and the surrounding inside wall surface will result in glare. For this reason, when small windows are used on the external walls, they must be designed with care. One traditional method of overcoming this problem is to use vertical slit windows which are usually located in the corners of rooms. Another method is to locate the windows between the ceiling and eye level, or alternatively filter can be used in the form of lattices, screens or shutters.

High humidity and typically overcast conditions in the warm humid regions result in a high proportion of the radiation being diffused so that the sky is the main source of glare. Because large openings are needed for cross ventilation, low overhanging eaves or wide verandahs are used to obstruct the view of most of the sky. In traditional houses thin external walls of coarsely woven mats, which in some cases can be rolled up, allow full advantage to be taken of every breeze.

3. Make statements saying when and where an architect would have to take into account the following (which times of the year and for which places):

a) excessive humidity

b) insufficient rainfall

c) excessive rainfall

d) excessive heat

4. Look at the following sketches of buildings from different parts of the world, and compare the buildings below from the following points of view:

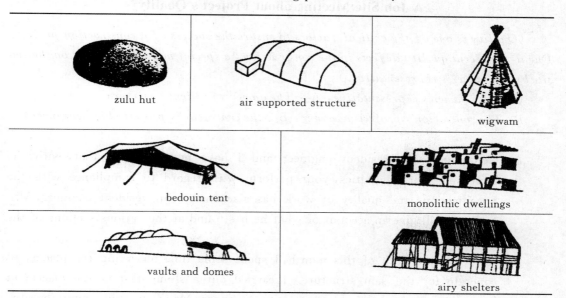

a) different forms due to climate
b) different types of structure
c) different materials used for their construction
d) different uses of the buildings
e) different social organizations and civilizations which use these buildings

Part 4　Business Activities

1. 文化与指南(Culture and Directions)

竣工工程质量的高低直接关系到工程的预期目的(the intended goal)是否能达到，投产后工程的效益是否能得到最大程度的发挥，因此，在工程的实施阶段，业主往往派遣自己的技术人员和管理人员到现场的各工作面去监督管理承包商的工程作业。

对于一个国际承包商(international contractor)，承建工程质量的优劣直接影响到他在国际工程承包界的信誉(reputation in the international contracting industry)。在实施工程的过程中，承包商要处理好质量、进度和费用之间的关系。越是在工期紧的情况下，越不能忽视质量。否则，很可能欲速则不达(The more haste, the less speed)。自始至终，承包商要按照业主颁发的图纸和其他技术文件施工，若发现其中的确有误，要及时提出，并提出修改意见。一旦出现质量问题，要及时改正。总之，对于一个国际承包商来说，一个优质的工程就是其最好的广告(A high-quality product is the best advertisement for an international contractor)。

2. 情景会话(Situational Conversation)

A Job Site Meeting about Project's Quality

(*Quality is one of the critical factors to ensure the success of a construction project. Due to the recent quality defects in the construction, a special meeting is called on the job site by the owner's representative.*)

(A: *The owner's representative*, A₂: *The owner's resident engineer*;

B: *The contractor's project manager*, B₂: *the contractor's project chief engineer*)

A: Gentlemen, our resident engineer and I have noticed lately that, with the increasing work activities, your perform with regard to compliance with the Specifications and quality of work has slackened. Our resident engineer, Mr. Jackson, will give an account of what he has found at the various sections of the job site.

A₂: On 18th and 19th of this month, I spent some time observing the placing of concrete for the dam structure. I carried out a slump test at the placement location with the help of your testing engineer, Mr. Zhou, and found that the water/cement ratio was not in accordance with the Specification. At the Power House, severe formwork misalignment has occurred along the draft tube side. This results in breaks and steps in the concrete. Areas of "honeycomb" can also be seen in the concrete due to lack of sufficient compaction and vibration. Also at the Power House, the multiple runs of embedded electrical conduit are being installed without adequate spacing. The space between such conduits must be at least equal to the largest aggregate size to allow proper bonding and to avoid forming a weakened plane in the concrete.

A: I'd like to have your comments on all these, Mr. Lin.

B: As the contractor's project manager, I believe quality is the life of the whole works and our first priority in its achievement. Unfortunately, such problems which Mr. Jackson have described have occurred. Regarding the problems, let's ask our chief engineer, Mr. Ma, to give the details of our remedy.

B₂: I was given the test result right after it came out. We immediately suspended the pouring operation and cleared the unqualified mix away. Mr. Jackson was then there. At the same time, our technicians looked into the cause and found that the electronic control system of the hatch plant was out of order. They soon repaired it and the pouring operation was resumed after a stop of 45 minutes. Now, about the misalignment. We've decided, after a careful study, that the defective concrete resulting from the formwork misalignment will be broken out with pneumatic drills. As for the areas of "honeycomb", we will get them specially tread to your satisfaction.

A: When do you plan to do the repairs?

B: Tomorrow.

A: What are you going to do about the last problem?

B: Actually, we are doing that job strictly in compliance with the working drawings issued by your engineer. May I refer you to Drawing MJ—7?

B_2: There's something in what Mr. Jackson has just said about the spacing between the conduits, but to my construction experience, weakened planes can be avoided if a special concrete mix is made and the pouring procedure is kept under proper control.

A: I will check the drawing and discuss the matter with our consultants this afternoon. Meanwhile, you are instructed to stop the work on this part of the project.

B: All right, we will take your instruction, but will you please give us a written confirmation?

A: OK. It will be given to you soon after this meeting.

B: Well, we will take a lesson from these matters and strengthen our quality assurance system.

A: I have to reiterate that the ongoing construction work is at a critical stage and that the power house is a key component of the generating capability of this project. I expect your site management to exert close supervision and proper coordination of the various work crews.

A_2: I also want to draw your attention to the water curing of concrete and storage of construction materials, especially the reinforcing steel.

B_2: We will pay more attention to those activities.

B: It's true that we are facing a pressing completion time and have to expedite the construction progress, but we will certainly not do it at the cost of the project's quality. I believe that efforts on both sides will result in not only a timely product but also a high-quality one.

New Words and Expressions:

critical factor	关键因素
quality defect	质量缺陷
call a meeting	召开会议
work activity	施工作业
compliance with the Specifications	按照规范
water/cement ratio	水灰比
misalignment n.	未对准,定线不准
draft tube	通风管,尾水管
breaks and steps	裂缝与凸面

honeycomb n.	蜂窝
compaction n.	压实,夯实
vibration n.	振动,振捣
embedded electrical conduit	电缆预埋管,电线预埋管
spacing a.	间距
bonding a.	结合,连接
weakened plane	弱结合面,弱结合层
first priority	第一优先权,首要任务
remedy v.	补救
pouring operation	浇注作业
unqualified mix	不合格的拌和料
electronic control system	电子控制系统
defective concrete	有缺陷的混凝土
break out	崩落,打掉
pneumatic drills	风钻
minor a.	微小的,不严重的
written confirmation	书面确认
take a lesson from	从……吸取教训
quality assurance system	质量保证体系
ongoing a.	进行中的
at a critical stage	处于关键阶段
generating capacity	发电能力
exert v.	施加,尽力
close supervision	严格监督
compromising on quality	对质量降低要求,使质量遭受损害
the curing of concrete	混凝土养护
a pressing completion time	急迫的竣工时间,工期紧

Notes to the Text:

1. resident engineer:驻地工程师,为业主方负责管理承包商实施工程的管理人员。

2. with regard to:有关。同义词组还有:in regard to, with respect to, concerning。

3. slacken:放松,松懈,形容词形式为 slack。

4. slump test:坍落度试验,也可以说 slump consistency test。slump:坍塌,坍落(度),工程中常见的术语还有:slump block(坍落试验堆),slump cone(坍落度筒),slump mete(坍落度计)。

5. multiple runs:管网布置。run 在这里是"线路"的意思,如:cable run(电缆敷设路线),pipe run(管道)。

6. look into the cause:调查原因。look into 在这里是"调查"的意思,等于 investigate,例如:look into a problem(调查一个问题)。

UNIT 12 THERMAL AND MOISTURE PROTECTION

7. resume：恢复，重新开始。这个词指的是事情暂停一段后又继续，如：resume the work(复工)，resume the talk(恢复谈判)，resume the journey(继续赶路)。

8. keep under proper control：恰当控制。under 这个词后面接名词构成一些很有用的词组，如：under discussion(在讨论中)，under consideration(在考虑中)，under repair（在修理中）。

9. working drawings：施工详图。

10. a timely product：直译是"一个及时的产品"，这里指的是"及时完成的工程"。

Post-Reading Exercises：

Now try to do the following translation：

1. "Actually, we are doing that job strictly in compliance with the working drawings issued by your engineer. May I refer you to Drawing No. MJ-7?"(事实上,我们是严格按照你方工程师颁发的图纸来做这项工作的。请你们看一下 MJ-7 号图纸好吗?)

"refer sb. to sth."："请某人参看某文件"。

1) A：Why have you reduced the progress?

B：I think we have the right to do so when your payment to us is delayed. ＿＿＿＿＿.
（请你看一下合同第二十条的规定好吗?）(Clause；Contract)

2) A：You have no right to change the design.

B：We have done that upon your approval. ＿＿＿＿＿.
（请你看一下 8 月 9 日的会议记录好吗?）(minutes of the meeting)

3) A：This section of the project should have been completed by the end of April. B：
＿＿＿＿＿＿＿＿＿＿＿＿＿＿＿＿＿＿＿＿＿＿＿＿＿＿＿＿＿.
（请你看一看你方的驻地工程师 7 月 8 日签发的停工令好吗?）(stop-work order; resident engineer)

2. "There's something in what Mr. Harrison has just said about the spacing between the conduits..."(Harrison 先生刚才关于预埋管之间的间距的看法有一定的道理……)

"There's something in what one says"："某人所说的话有一定的道理"。

1) A：The building will look better if glass curtain walling is adopted.

B：＿＿＿＿＿＿＿＿＿＿＿＿＿＿＿＿＿＿＿＿＿.
（你说的有一定道理,但增加的费用从哪里来呢?）

2) A：A lot of money will be saved if the design is modified a little.

B：＿＿＿＿＿＿＿＿＿＿＿＿＿＿＿＿＿＿＿＿＿.
（你说的有一定道理,但是这是否会影响整个设施的功能?）(facility)

3) A：If you recruit local laborers, it will relieve your manpower shortage.

B：＿＿＿＿＿＿＿＿＿＿＿＿＿＿＿＿＿＿＿＿＿.
（你说的有一定道理,但他们具有所要求的技能吗?）

3. "Meanwhile, you're instructed to stop the work on this part of the project."(同时,现指示你方停止这一部分工程的工作。)

"You're instructed to do sth." is a formal expression, meaning"现指示你做某事"。

1) 现指示你方暂停发货。(suspend)

2) 现指示你方停止立模。(erect, formwork)

3) 现指示你方复工。(resume)

Part 5 Extensive Reading

(A) Design for Different Climates

The shape of the house is strongly influenced by the climate. Where it is warm, the plan of the house is open, with the rooms often arranged round a courtyard which admits air but not too much sun. In the north, houses are more compact so that they can be more easily keep warm in winter; where there is much rain, they have steep roofs to throw it off; but where there is much snow and frost, as in Switzerland, they generally have flatter roofs on which the snow will lie, making a warm blanket over tie house. In hot countries, flat roofs are common because no slope is needed to throw off rainwater and because a flat roof is for sleeping in the hottest weather.

The shape of windows is also dependent on the climate. They are large in the north to admit sunlight, though not so large as to make the rooms too cold. In the south, windows are small so as to keep the house as cool as possible inside, and are often shaded from the direct glare of the sun by balconies or verandas which provide a cool sitting place in the open air. Shutters outside the windows also provide protection from the sun. Windows are placed facing away from the sun in hot countries and where possible, towards the sun in cold climates to let in as much light and warmth as possible. Chimneys are a prominent feature of the exterior of the northern house.

The materials of which houses are built play a large part in giving character to the scenery of different countries. In England, before modern transport made it possible to carry cheap bricks all over the country, and before standardized building materials were made in factories, every region had its characteristic building material. Because old houses are built of local materials, they fit into the landscape, and their color and texture harmonize with it. Efforts are still made, therefore, to build as far as possible in local materials, especially in country distracts.

The shape and size of the rooms and therefore of the house as a whole, depend on the way people live. In the Middle Ages, when people spent most of their time out of doors, rooms were few and barely furnished. But as indoor activities increased, there was more emphasis on indoor comfort, and rooms were set apart for different purpose. Nowadays in the West, the desire for privacy has led to small houses or flats with small rooms, so that each family can have a separate living place and each person a separate room. In warm countries people live much more out of doors than in the north, and consequently, the houses are simpler and more barely furnished. In Japanese houses, the dimensions of all the room s are based on those of the mats with which all the floors are covered. The mat is always of the same size, so that each room is so many mats wide and so many long, thus making all house consistent in scale and proportion.

Notes to the Text:

1. The materials of which houses are built play a large giving character to the scenery of different countries: 建造房屋用的材料，在表现各个国家的风光特色中起了很大的作用。

2. In Japanese houses, the dimensions of all the rooms are based on those of the mats with which all the floors are covered: 在日本式房屋里，所有房间的面积是按照铺在整个地板上的席子的大小而设定的。

3. ... so that each room is so many mats wide and so many long, thus making all houses consistent in scale and proportion: 所以，每一间房间的宽度是若干块地席的宽度，长度也是同样的长度，这样就使得所有房间在规模和比例上是一致的。

New Words and Expressions:

compact	*a.*	紧密的
steep	*a.*	陡峭的
frost	*n.*	冰冻，霜
Switzerland	*n.*	瑞士
slope	*n.*	倾斜，坡度，斜面
glare	*n.*	强烈的光
balcony	*n.*	阳台
veranda	*n.*	走廊，阳台
prominent	*n.*	突出的
feature	*n.*	特征
exterior	*n.*	外部
northern	*a.*	北部的，朝北的
scenery	*n.*	风景，景色
local	*a.*	当地的，地方的
texture	*n.*	（材料的）构造，构成，织物
landscape	*n.*	风景，景色
harmonize	*v.*	与……协调
country district		郊区
furnish	*v.*	装备，家具；布置（房间）

privacy　　　　n.　　　　　独处，隐居
consequently　ad.　　　　　因而，所以
dimension　　 n.　　　　　尺寸，大小(pl. 面积)
mat　　　　　 n.　　　　　席子，地席
consistent　　a.　　　　　一致的，连贯的

Post-Reading Exercises:

1. Translate the following expressions:
1) the dimension of the mats
2) the windows facing towards the sun
3) the prominent feature of Japanese houses
4) to harmonize with the local landscape
5) to make bedrooms have as much sunlight as possible
6) the character of Chinese ancient buildings
7) social customs
8) to suit the appearance of houses to the local conditions

2. Fill in the blanks with the proper words given in the box:

| texture | local | flat | glare | shade | landscape | include |
| shutter | dwelling | courtyard | balcony | make out | veranda |

1) While he was leaning against the balusters of the _____ outside his bedroom looking at the flowers in the _____, one of the nails happed to break and he nearly fell off.

2) My friend told me that his newly-bought _____ was quite spacious, which consist of five rooms not _____ the kitchen.

3) She pulled the _____ outside the window so as to prevent the direct _____ of the setting sun.

4) My sister _____ her eyes from the sun in order that she could _____ the man who was walking towards us.

5) As the houses are built of _____ materials, they fit into the surrounding _____; their color and _____ make a whole with it.

6) I found the _____ there were quite simple and crude.

7) A _____ is a porch or portico along the outside of a building, which is sometimes partly enclosed.

(B) Thermal Insulation

Thermal insulation is any material that is added to a building assembly for the purpose of slowing the conduction of heat through that assembly. Insulation is almost always installed in new roof and wall assemblies in North America, and often in floors, around foundations and concrete slabs on grade. To say it shortly, nearly anywhere that heated or

UNIT 12 THERMAL AND MOISTURE PROTECTION

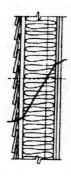

wall assembly　　　　　　　roof assembly
Temperature differences in roof and wall assemblies

cooled interior space comes in contact with unheated space, the outdoors or the earth.

The effectiveness of a building assembly in resisting the conduction of heat is expressed in terms of its thermal resistance, which is abbreviated as "R". The thermal resistance R is expressed either in English as British Thermal Units (BTU) or in metric units of square meter-degree Celsius per Watt. The higher the R-value, the higher the insulating value.

Every component of a building assembly contributes to its overall thermal resistance. The amount of the contribution depends on the amount and type of materials. Metals have very low R-values, and concrete and masonry materials are only slightly better. Wood has a substantially higher thermal resistance, but not nearly as high as that of common insulating materials. Most of the thermal resistance of any insulated building assembly is attributable to the insulating material.

In wintertime, inside a building it is warm, but outside it is cold. The inside surface of a wall is warm, and the outside surface is cold. Between the two surfaces, there is a big gap in temperature. The indoor temperature and the outdoor temperature vary according to the thermal resistances of the various layers of the assembly. The largest temperature difference within the assembly is between the inner and outer surfaces of the insulation.

Thermal insulation for a low-slope roof may be installed in any of 3 positions: below the structural deck, between the deck and the membrane or above it.

For insulation below the deck, the insulation is installed above a vapor retarder, either between framing or on top of a suspended ceiling assembly. A ventilated air space should be provided between the insulation and the deck. Insulation in this position is relatively economical and trouble-free, but it leaves both the deck and the membrane exposed to the full range of outdoor temperature fluctuations.

The traditional position for low-slope roof insulation is between the deck and the membrane. Insulation in this position must be in the form of low density rigid panels or lightweight concrete in order to support the membrane. The insulation protects the deck from temperature extremes and is itself protected from the weather by the membrane.

The protected membrane roof insulation above the roof membrane is a relatively new concept. It offers two major advantages: the membrane is protected from extremes of heat and cold, and the membrane is installed on the warm side of the insulation, where it is immune to vapor blistering problems. Because the insulation itself is exposed to water when placed above the membrane, the insulating material must be one that retains its insulating value when wet and does not decay or disintegrate.

New Words and Expressions:

thermal insulation	保温隔热
building assembly	构造体
install v.	安装
abbreviate v.	缩写
Celsius a.	摄氏的
retard v.	阻碍,迟滞
decay v.	腐烂
disintegrate v.	分解

Post-Reading Exercises:

1. Decide whether the following statements are true(T) or false(F) according to the passage given above.

1) A thermal insulation is the part of a wall or a roof that helps to reduce the conduction of heat through that assembly.   (   )

2) The higher the R-value, the higher the insulating value.   (   )

3) Metals have very low R-values, but concrete and masonry have high R-values.   (   )

4) During the summer, a building is warm inside and cold outside.   (   )

5) The inside surface of a wall or roof assembly is warm, and the outside surface is cold.   (   )

6) Usually, the insulation of low-slope roofs can be found above the roof membrane.   (   )

7) In insulations above the membrane, the membrane is protected from extreme heat and cold, which is a major advantage of this type.   (   )

2. Fill in the blanks with the proper words given in the box changing the form if necessary:

purpose	contribute	resistance	expose	extreme
immune	retain	suspend	attributable	component

1. A wise mother never _____ her children to the slightest possibility of danger.

2. According to the statistics, a third of the citizens of many civilized countries admit to suffering from _____ loneliness.

UNIT 12   THERMAL AND MOISTURE PROTECTION

3. The criminal was told he would be _____ from punishment if he helped the police.
4. We climbed to the top of the mountain with the _____ of watching sunrise.
5. However, people _____ diverse attitudes towards these new measures.
6. His untimely death was _____ in part to overwork and lack of exercise.
7. It's time to set our differences by and work together for a common _____ .
8. The report says design faults in both the vessels _____ to the tragedy.
9. Where there is oppression, there is _____ .
10. Mongolian medicine is the important _____ section of traditional medicine.

UNIT 13   PROPORTIONS IN DESIGN

Part 1   Warm-up Activities

1. Look and read:

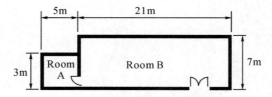

floor plan of a building

The *ratio between* the length *and* width of Room A *is* 5 : 3 (five to three).
The *ratio between* the length *and* width of Room B *is* 3 : 1 (three to one).
Room B *is wider than* Room A, but its width is *less in proportion to* its length.
Therefore Room B is *relatively narrow* or *proportionately narrower*.

2. Now look at these diagrams showing the relation between size and supporting strength and answer the questions:

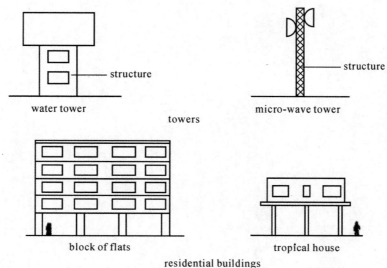

UNIT 13   PROPORTIONS IN DESIGN

Questions:

a) Which tower carries a relatively heavy load?

b) Which building carries a relatively light load?

c) Which part of the block of flats supports its weight?

d) Which part of the tower supports its weight?

e) What is the approximate ratio between the length of the columns of the block of flats and the height of the building?

f) What is the approximate ratio between the length of the columns of the tropical house and the height of the building?

g) Which building has longer columns in proportion to its size?

h) What is the approximate ratio between the length and thickness of the columns of the block of flats? (This ratio is called the slenderness ratio.)

i) What is the approximate ratio between the length and thickness of the columns of the tropical house?

j) Which building has proportionately thicker columns?

3. Make sentences from this table:

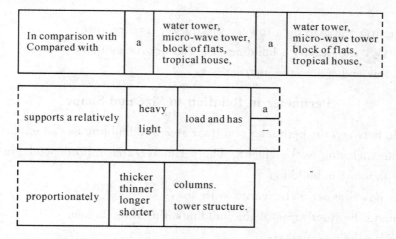

4. Now read these two paragraphs and add the missing words:

a) If we *compare* the columns supporting the two buildings, *we can see* that the columns of the block of flats are *relatively* short and thick *in proportion* to its size, *while* those of the tropical houses _____. We *can conclude* that the heavier building needs *proportionately* shorter and thicker columns, *whereas* _____.

b) *The explanation for this is that* short thick columns are stronger than long and thin ones *since* the strength of the column *depends on* its thickness and its length. Supporting strength is *directly* proportional to _____ and *inversely* proportional to _____. *Consequently, the heavier* the building, *the* _____ and _____ its columns, and *conversely*, the lighter the building _____.

201

Part 2 Controlled Practices

1. Decide whether these statements are true or false and try to correct the false ones.

a) The ratio between the height and width of the micro-wave tower is higher than that between the height and width of the water tower. (1 : 3 is a higher ratio than 1 : 2).
(   )

b) The structure of the water tower has to support less weight than that of the micro-wave tower.
(   )

c) The columns of the block of flats have greater supporting strength than those of the tropical house.
(   )

d) The strength of a column is directly proportional to its height and inversely proportional to its thickness.

e) Compare with a micro-wave tower, a water tower has a relatively tall structure.
(   )

f) The lighter the load on a tower, the thicker its structure.   (   )

g) Similarly, the heavier a building, the thinner its columns.   (   )

2. Read this and follow the instructions:

Perimeter in Relation to Size and Shape

The ratio between the perimeter and floor area of a building has an important effect on the cost of the enclosing wall element. The perimeter/area ratio depends on the size and shape of the plan of the building.

To show how the perimeter varies with size:

1) Calculate the floor areas of the buildings illustrated below.

2) Calculate their perimeters.

Find the ratio between the perimeter and the floor area for each building.

building A   building B

a) Floor area= _____   d) Floor area= _____
b) Perimeter= _____   e) Perimeter= _____
c) Perimeter/area ratio= _____   f) Perimeter/area ratio= _____

Now complete these statements by choosing from the words given in the bracket:

g) By comparing the ratio of perimeter to floor area for the two buildings we can see that the _____ (larger/smaller) building has higher perimeter/area ratio.

h) We can conclude that smaller buildings have a _____ (longer/shorter) perimeter in proportion to floor area than larger buildings.

3. Now read and complete these:

To show how perimeter also varies with shape:

These floor plans have the same area but they differ in shape. Do they have the same perimeter? Calculate the perimeters of the square and rectangular buildings.

Shape	Floor area	Perimeter
a) The circle	196 m²	49.6 m
The square	196 m²	_____
The rectangle	196 m²	_____

b) The circular building, which has the most compact shape, has the smallest perimeter in proportion to area, whereas the _____, which has the least _____, has the _____ perimeter in proportion to area.

c) If we _____ the perimeters of buildings with the same floor area but different shapes, we will _____ that the more compact the shape, _____ _____.

d) We can _____ that _____ ratio depends on _____ as well as _____.

Part 3 Further Development

1. Read the paragraphs and look at the diagrams:

The effects of the surface area/volume ratio in architecture:

The relation between surface area and volume has many effects on the performance of buildings. For example, the rate at which a building gains or loses heat through its walls depends on its surface area/volume ratio. Heat transfer is directly proportional to surface area and inversely proportional to volume. Thus a building with a proportionately large surface area, such as a one room house, will lose or gain heat relatively rapidly. Conversely, a building with a large volume in relation to its surface area, such as a block of flats, will retain more heat.

Heat losses from a building are reduced by using insulating materials such as expanded polystyrene. Thickness of insulation is in inverse proportion to heat transfer.

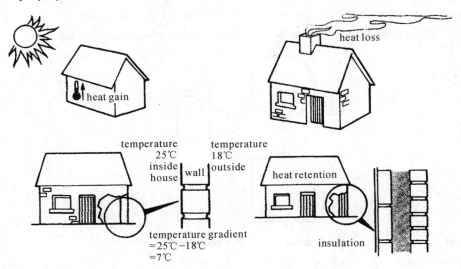

2. Make true statements from these tables:

a)

Heat transfer Heat loss Heat gain Heat retention	is	directly inversely	proportional to	air temperature gradient. thickness of insulation. surface area. volume.

b)

The higher the ratio between surface area and volume,	the more quickly it regains or loses heat.
The lower the ration between surface area and volume,	the more it retains heat.
The smaller the size of the building,	the fast the rate of heat transfer.
The larger the size of the building,	the less it retains heat.
The thicker the insulation of a building,	the more slowly it gains or loses heat.
The less compact the shape of a building,	the slower the rate of heat transfer.

3. Read and think:

We can conclude that the more compact the shape of the plan of a building for a given area, the less the heat loss. It can also be shown that for a given required total floor area in a two or more storey building, the higher the building, the greater the heat loss. However, buildings gain heat from the sun as well as losing heat to the cold. The more direct the face of a building to the sun, the greater heat it gains.

UNIT 13 PROPORTIONS IN DESIGN

The following examples illustrate the effects of the perimeter/area ratio, the surface area/volume ratio and orientation of the building on heat transfer. Explain them by answering the questions:

a) Igloos are built by Eskimos in the Arctic where the cold is very intense. Why do they build them this shape?

b) Mud houses are built by people in the tropics where the heat is very intense. Why do they build the walls so thick?

c) Why are some houses in tropical climates built with plans shaped like this?

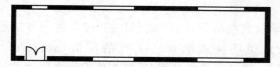

d) These two buildings have identical floor areas. Which of them loses the greater amount of heat? Why?

e) Why do radiators have fins?

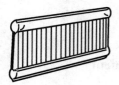

f) Which of these two south-facing elevations will receive the greater amount of solar radiation? Why? Will the solar radiation be greater in summer or winter?

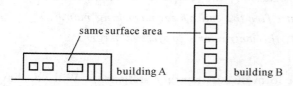

g) Both these buildings have the same floor area and the same height. Which one will

205

be more expensive to heat?

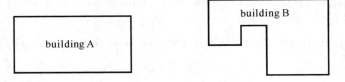

Part 4　Business Activities

Payments

1. 背景与指南(Background and Directions)

工程款项的支付贯穿着工程的整个实施过程,付款程序(payment procedure)也比较复杂。按照土木工程施工合同的支付惯例,分为进度款和最终结算款(interim and final payment)。

进度款按月支付,其程序如下:首先,承包商在每个月末(或下月初)按业主一方指定的格式提交每月报表(monthly statement),说明完成的永久工程的价值(value of the permanent works);相关临时工程、施工设备、计日工等的费用;运至工地但尚未使用的材料和永久工程设备的发票价值(invoice value);价格的调整额以及业主同意的索赔款等。当业主代表或业主工程师收到报表后,认为正确无误,在扣去保留金(retention money)和其他应扣款后,应在合同规定的时间内开具支付证书(payment certificate)并送交业主。收到支付证书后,业主应在规定的时间内进行审查并向承包商支付进度款。

最终结算款大体程序如下:当工程竣工,业主应退还一半保留金,同时承包商向业主提交竣工报表(statement at completion),说明他完成的所有工作的价值以及业主应付给他到期的工程款余额(balance due),业主根据合同规定予以支付。当维修期结束后,业主须在规定的时间内退还另一半保留金。

2. 情景会话(Situational Conversation)

(1) Settling the Form of the Monthly Statement

(*For a construction project, payment of the contract price, normally, is divided into interim payments and final payment, and is rather complex. To receive the progress payments during the period of the construction, the contractor is required to apply for them by submitting a monthly statement after the end of each working month.*)

(A: *the owner's representative*; B: *the contractor's project manager*)

B: Mr. Madison, this is the draft form of our monthly statement. Will you have a check to see if it is all right, please?

A: I hope so too. I think I'll be finished in fifteen minutes and then you can have my comments right away. Would you like a coffee?

B: Thank you.

A: (after checking) Well, Mr. Zhang, the form is OK in general, but Item 1 "Value of the Works Executed" in this form should be broken down into two items. One is for the permanent works and the other for the temporary works.

B: Thank you, Mr. Madison. I'll come back tomorrow afternoon and give you the Monthly Statement for the completed month in this revised form. Is it all right with you?

A: Anytime after 3 o'clock.

B: When do you think we can receive the payment?

A: If there's no problem with the statement, I'll certify the payment to our head office and accounting department right away. It will be transferred to your bank account probably within 30 days after I get the statement, or within 56 days at the latest.

B: OK. I'll be expecting it then. The sooner, the better.

(2) Checking Payment

(As the amount of the payment certificate is obtained from complex calculations, problems may occur easily. Mr. Zhang is in the owner's site office, showing his doubt to Mr. Madison about the amount to be received from the owner in respect of the monthly payment.)

A: Have you received the payment for October, Mr. Zhang?

B: I am not sure. We have not been informed by the bank yet, but thank you for your prompt delivery of the copy of the Interim Certificate No. 6. Unfortunately, we found that the amount is incorrect.

A: Oh, is it?

B: In this certificate, an amount of 10 percent of the payment due to us has been deducted to repay the advance payment. We don't think that it should be.

A: The owner is entitled to repayment of the advance through deductions at a rate of 10 percent of each progress payment.

B: Yes, but after the total amount certified has reached 30 percent of the contract price, according to the contract.

A: Now, the payments you've got have added up to this amount.

B: That's right, but we have a different understanding on this point. We believe that the contract implies that the deduction should start from the next certificate by using the word "AFTER" in Clause 60 "Repayment of the Advance Payment".

A: Perhaps the word "AFTER" is used incorrectly. It is indeed quite misleading here, but the interim certificate has been delivered to our accounting department back in our office. Let me put it this way: the amount in this certificate stays as it is and I'll make no deduction for the repayment in the next certificate. How does this sound to you?

B: It sounds all right. I take it, since it would require too much effort to modify the amount of the current certificate. Good cooperation is the key for the success of this project.

A: Thank you for your understanding.

B: Mr. Madison, there's something else I want to talk over with you. It's about the retention money. By now, the retention money that has been deducted from all the previous certificates has reached half of the total. Is it possible for us to substitute a bank guarantee for the money you have retained? That means we submit to you a guarantee with the same amount and you release all the retained money to us. We're in need of more working capital at this construction stage, you know.

A: This concerns our general financing arrangements. Personally, I don't oppose such an idea so long as the guarantee is an unconditional one and the issuing bank is acceptable. This issue, however, is beyond my authority and I'll have to ask the head office for approval.

B: Will you please advise us when you have a response from your head office?

A: I'll give you a reply in seven days.

B: Thank you.

New Words and Expressions:

contract price	合同价格
final payment	结算
progress payment	进度付款
settle the form	将格式定下来
permanent works	永久工程
temporary works	临时工程
revised form	修改的格式
payment certificate	支付证书
in respect of	有关……方面
show one's doubt about	对……表示疑问
the advance payment	预付款
reach 30 percent of the contract price	达到合同价格的30%
retention n.	保持,保留(金)
working capital	流动资金,周转资金
unconditional a.	无条件的

issuing bank					开具保函的银行，开证行

Notes to the Text：

1. interim payment：期中付款，临时付款。相当于我们通常所说的"进度付款"(progress payment)。

2. monthly statement：每月报表。指承包商为获得项款，在每月末或下月初向业主提交的报表，其中要说明在该月已完成的工程价值、购买并运至现场的材料和设备的价值(value of the materials and plants to be delivered onto the site)以及有关的价格调整(price adjustments)等。

3. Item 1 "Value of the Works Executed"：第一项"完成的工程的价值"。item 这个词在工程管理中很常用，如：items of work（工作项），an item of equipment（一件设备），item number（项目编号）。

4. break down into two items：分成两项。break down 在这里是"分解"、"分成细项"的意思如：break down the prices（价格分解，价格分析）。

5. certify the payment：开具支付证书。certify 在这里的意思是"证明"，"为……开具证书"，常用于财务方面，如：certified invoice（签证发票），certified check（保付支票），certified letter of credit（保付信用证）。

6. interim certificate：即 interim payment certificate（期中支付证书）。

7. the advance：即 the advance payment（预付款），指业主在承包商开工前提前支付给他的一笔不计利息的款项，用于承包商的前期开支，并按合同规定从每月进度付款中分期扣还。

8. stay as it is：保持不变，不做修改。这是一种习惯用法。

9. the current certificate：这一期的证书。current 意思是"现时的"，"通用的"，如：current affairs（时事），current account（活期存款账户，经常项目），current assets（流动资产），current income（本期收益），current deposit（活期存款），current liabilities（流动负债，短期负债），current price（时价）。

10. How does this sound to you? 你觉得这个意见怎么样？这是口语中征求别人意见时的常用句型。

11. retention money：保留金。指业主为了保证承包商在维修期/缺陷责任期(maintenance period/defects liability period)内去修补工程缺陷而从进度款项中所暂时扣发的一笔款额，约为合同价格的5%。

12. substitute for：用……取代……。注意这个词组与 replace with 的区别，例如：The engineer substituted the new parts for the worn parts of the machine.（那位工程师用新部件换下了机器的磨损件。）这句话同样可以说：The engineer replaced the worn parts of the machine with the new parts.

13. bank guarantee：银行保函。在国际工程管理中，常用的银行保函类型有：bid guarantee（投标银行保函），performance guarantee（履约银行保函），bank guarantee for advance payment（预付款银行保函），domestic preference guarantee（国内优惠银行保函）。

Post-Reading Exercises:

Now try to do the following translation:

1. "I'd prefer to wait here if you don't mind."(如果你不介意的话,我在这里等一下。)"...if you don't mind" is a very polite way in making a suggestion or a choice.

 1) A: We'd like to invite you to a dinner party tonight.

 B: _____.

(如果你不介意的话,我想待在我的房间准备明天的谈判文件。)(negotiation; document)

 2) A: Can we settle the matter now?

 B: _____.

(如果你不介意,我想把这一问题留在下次会议讨论。)

 3) A: Where would you like to go for a visit?

 B: _____.

(我想逛一逛故宫,如果你不介意的话。)

2. "I think I'll be finished in fifteen minutes."(我想我十五分钟就能完成。)"be finished (with sth.)" means "(已经)完成某事。"

 1) A: _____?(你的工作完成了吗?)

 B: Not yet.

 2) A: _____?

(你认为你们何时能完成截流?)(river closure)

 B: We are conducting the job right now and will be finished in three days.

Part 5 Extensive Reading

(A) Shape Effect on Cost

  The shape of a building has an important effect on cost. As a general rule, the simpler the shape of the building the lower will be its unit cost. As a building becomes longer and narrower, or its outline is made more complicated and irregular, so the perimeter/floor area ratio will increase accompanied by a higher unit cost. The significance of perimeter/floor area relationships will be considered in more detail in the unit. An irregular outline will also result in increased costs for other reasons. For example, site works and drainage work are all likely to be more complicated and more expensive. The additional costs do not finish there, as brickwork and roofing will also be more costly due to the work being more complicated. It is important that both architect and client are aware of the probable additional costs arising from comparatively small changes in the shape of a building. (A)

Although the simplest plan shape, that is, a square building, will be the most economical to construct, it would not always be a practicable proposition. In dwellings, smaller offices, schools and hospitals, considerable importance is attached to the desirability of securing adequate natural day lighting to most parts of the buildings. A large square structure would contain areas in the center of the building, which would be deficient in natural lighting. Difficulties could also arise in the planning and internal layout of the accommodation. Hence, although a rectangular shaped building would be more expensive than a square one with the same floor area because of the smaller perimeter/floor area ratio, nevertheless practical or functional aspects and possibly aesthetic ones in addition, may dictate the provision of a rectangular building. (B)

Certain types of building present their own peculiar problems, which in their turn may dictate the form and shape of the building. For instance, hotels to provide guests with good views and the advertising effect of a prominent building on the skyline, need to be tall. The shape and floor area are closely related to the most economic bedroom per floor ratio and this is generally in the range of forty to fifty. This dictates a tall slab rather than a tower. Slender towers are aesthetically desirable but their relatively poor ratio of usable to gross floor area often renders them expensive. (C)

The shape of a building may also be influenced by the manner in which it is going to be used. For instance, in factory buildings the determining factors may be co-ordination of manufacturing processes and the form of the machines and the finished products. In schools, dwellings and hospitals, shape is influenced considerably by the need to obtain natural lighting. Where the majority of rooms are to rely on natural lighting in daylight hours? The depth of the building is thereby restricted. Otherwise it is necessary to compensate for the increase in depth of building by installing taller windows, which may compel increased storey heights. The aim in these circumstances should be to secure an ideal balanced solution, which takes into account both the lighting factor and the constructional costs. (D)

New Words and Expressions:

perimeter	n.	周长,周边
setting out		设计；装饰,布置
drainage	n.	排水(设备),下水道
brickwork	n.	砌砖；砖房
roofing	n.	屋面工程
proposition	n.	提议；主张
desirability	n.	可取之处；好处,优点
deficient	a.	不足的；缺乏的
rectangular	a.	长方形的,短形的
aesthetical	a.	美学的；美的,艺术的

slab *n.*　　　　　　　　　　平板；厚片，厚板
compensate for　　　　　　　补偿

Post-Reading Exercises：

1. Choose the one that best completes the sentence.

1) It can be inferred from the text that _____.

A. a rectangular shaped building has a smaller perimeter/floor area ratio.

B. a large square building needs extra lighting systems

C. a rectangular building is preferable in any case

D. a tower provides guests with better views

2) Why should hotels be tall? _____

A. To draw attention

B. To dwarf surrounding buildings

C. To provide billboards

D. To add color to the sky

3) In designing factory buildings, which of the following should be considered first? _____

A. The provision of natural lighting.

B. A artificial ventilation to internal rooms.

C. Aesthetic aspects.

D. Manufacturing processes and forms of facilities and final products.

4) The author implies that taller rooms are accompanied by _____.

A. a decrease in construction maintenance and heating costs

B. increased depth of the building

C. consumption of less electricity

D. elongated corridors

2. Match the main for each paragraph with their appropriate paragraph numbers A—D.

1) The shape of a building has an important effect on expenses.　　　　　　　　(　　)

2) The way in which the building is going to be used may influence its shape.

(　　)

3) In terms of shape and form the peculiar problems with a building must be taken into consideration.　　　　　　　　(　　)

4) A square building, though the most economical to construct, is not always preferable.　　　　　　　　(　　)

(B) Future Trends in Construction

The future has many surprises in store for us. We are already thinking of possibilities which are still only a gleam in the experts' eyes.

Excavation will be performed for large projects by applying nuclear methods with special radiation-proof equipment and methods.

Prefabrication will be much more advanced and a universal standardized code based on using metric modular units will be applied on a global basis. Not only will prefabrication permit assembly of building components manufactured in different locations at distant sites, but it will also be engineered so that various appliances and equipment, which are equally standardized, can be used anywhere, built-in if so desired, and with no worry about not being able to find spare parts or repair facilities locally in case of mechanical failure.

Many daily functions will become automated and computer-controlled, especially in residential construction. Lighting, heating, kitchen services and many other functions will be timed to suit personal requirements and the complete integrated program will be capable of being triggered or adjusted by remote control commands given by telephone. General news, private messages, pictures and even games will be projected on special wall and ceiling panels by means of built-in projection equipment. Many other functions in business and industry will be treated similarly. Miniaturization of formerly bulky items, thanks to modern electronic know-how, including micro-circuitry and solid-state technology, will not only reduce size and space requirements of many building components control elements, but in view of the reduced power requirements will also help to conserve energy. All this will require special facilities and installations and will add greatly to the complexities of both design and construction.

Our changing energy needs and sources will affect building design greatly. Solar power will at first tend to be developed on an individual basis, with various houses or buildings having their own installations. This may be followed by complete solar power stations combined with underground hot water storage reservoirs and other ancillary facilities. Wind power will be utilized more in wind power stations. Coal will now be economically justified in comparison with the escalating costs of petrochemical and nuclear energy. Nuclear power stations may become commonplace for general power generation, but a lot of equipment will be powered by mini-nuclear power units, all mounted in accident-proof sealed units, which will outlast the equipment and be capable of being reused again and again.

Construction methods will change in all areas and manual labor will tend to disappear with more and more emphasis on mechanized operations. Thus there will be a demand for more specialists, primarily those having technical backgrounds.

Research and experimentation by inventive minds, both in the field and in the laboratory, will lead us into an era of progress in technology which will open new vistas and enable us to use our vast technical knowledge for the good of human beings throughout the world.

New Words and Expressions:

gleam	n.	微光
excavation	n.	挖土,挖方工程
engineer	vt.	设计,建造,制定
appliancen	n.	器具,设备,装置
residential	a.	居住的,住宅的
time	v.	使……合时宜
trigger	v.	触发,启动
miniaturization	n.	微型化
bulky	a.	笨重的,体积大的
micro-circuitry	n.	微型电路(技术)
conserve	v.	保存,节省
complexity	n.	复杂性,错综复杂
pariah	n.	受轻视者,为社会所遗弃者
ancillary	a.	辅助的,附属的
prominence	n.	著名,杰出
scrub	v.	洗涤,清洗
purification	n.	净化,提纯
ecologically	ad.	生态学上地
escalate	v.	日益提高
outlast	v.	较……经久(耐用),比……持续得久
vista	n.	远景,展望
in store for		必将发生
thanks to		由于
radiation-proof equipment		防辐射设备
metric modular units		米制模数单位
in comparison with		与……相比

Notes to the Text:

1. Not only will prefabrication permit assembly of building components manufactured in different locations at distant sites, but it will also be engineered so that various appliances and equipment, equally standardized, can be used anywhere, built-in if so desired, and with no worry about not being able to find spare parts or repair facilities locally in case of mechanical failure.

("Not only" introduces an inverted sentence, "so that" introduces a clause of purpose.)

预制业的发展不仅有可能在遥远的工地上装配不同地区预制的建筑构件,而且也可以设计采用完全标准化的各类设备和装置,可用于任何地方,如果需要还可以埋设在内,万一发生机械故障,不必为在当地找不到备用件或无法修理而发愁。

2. Miniaturization of formerly bulky items, thanks to modern electronic know-how, including micro-circuitry and solid-state technology, will not only reduce size and space requirements of many building components control elements, but in view of the reduced power requirements will also help to conserve energy.

(This is a simple sentence with two predicates. "thanks to modern electronic know-how" is a parenthesis.)

由于现代电子技术,包括微型电路和固态电路技术,以前笨重的设备趋于小型化,这不仅有助于缩小许多房屋构件控制部件的尺寸和空间要求,而且由于减少了电力需要,还有助于节省能源。

3. Solar power will at first tend to be developed on an individual basis, with various houses or buildings having their own installations.

("with various houses or buildings having their own installations" is an absolute construction of the present participle, used as an adverbial.)

太阳能首先趋向于在个别房屋或建筑物上开发利用,然后各种房屋或建筑物都有自己的太阳能设备。

4. Research and experimentation by inventive minds, both in the field and in the laboratory, will lead us into an era of progress in technology which will open new vistas and enable us to use our vast technical knowledge for the good of human beings throughout the world.

在施工现场和实验室里的创造性的研究和实验将引导我们进入一个技术进步的新时代,这个时代将展现出新的前景,使我们能用广博的技术知识造福于人类。

Post-Reading Exercises:

1. Translate the following sentences into Chinese:

1) The future has many surprises in store for us.

2) We are already thinking of possibilities which are still only a gleam in the experts' eyes.

3) Excavation will be performed for large projects by applying nuclear methods with special radiation-proof equipment and methods.

4) Prefabrication will be much more advanced and a universal standardized code based on using metric modular units will be applied on a global basis.

5) Many daily functions will become automated and computer-controlled, especially in residential construction.

6) Lighting, heating, kitchen services and many other functions will be timed to suit personal requirements and the complete integrated program will be capable of being triggered or adjusted by remote control commands given by telephone.

2. Decide whether the following statements are true(T) or false(F) according to the passage given above.

1) General news, private messages, pictures and even games will be projected on

special wall and ceiling panels by means of built-in projection equipment.　　　　（　）

2) Many other functions in business and industry will be treated differently.　（　）

3) All this will not necessarily require special facilities and installations and will add greatly to the complexities of both design and construction.　　　　　　　　　（　）

4) Our changing energy needs and sources will affect building design greatly.　（　）

5) Solar power will finally tend to be developed on an individual basis, with various houses or buildings having their own installations.　　　　　　　　　　　　　（　）

6) This may be followed by complete solar power stations combined with underground hot water storage reservoirs and other ancillary facilities.　　　　　（　）

7) Wind power will be utilized more in wind power stations. Coal will now be economically justified in comparison with the escalating costs of petrochemical and nuclear energy.　　　　　　　　　　　　　　　　　　　　　　　　　　　　　　　（　）

8) Construction methods will change in all areas and manual labor will tend to disappear with more and more emphasis on mechanized operations. Thus there will be a demand for more specialists, primarily those having technical backgrounds.　　（　）

UNIT 14   POST-CONSTRUCTION PROBLEMS

Part 1   Warm-up Activities

1. Look and read:

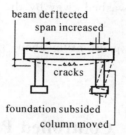

Example: The foundation subsided with the result that/and as a result, the column moved.

Now make similar cause/effect statements according to the table.

Cause	Effect
foundation subsided	column moved
column moved	span increased
span increased	beam deflected excessively
beam deflected excessively	cracks formed on the underside of beam

2. Look at these diagrams and put the events in the correct order to make cause/effect tables as in exercise 1.

    a) concrete floor expanded

    cracks formed in floor

    hardcore below the floor contained soluble salts

    salts interacted with cement in concrete floor

    b) roof expanded, wall/roof joint failed

    heavy rain washed away gravel on roof

    roof heated up

    roof inadequately protected from the sun

c) gaps formed between window and frame
woodwork expanded
moisture content of wood increased
wood was painted with poor quality paint
later the wood dried and contracted

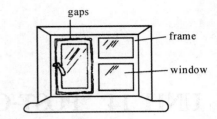

3. Read this:

Q: How did the subsidence of the foundation lead to/bring about cracks in the beam?

A: The subsidence of the foundation resulted in/caused the movement of the column. This, in turn, resulted in/caused an increase in the span of the beam, excessive deflection of the beam and the formation of cracks in the underside of the beam.

Now write answers to these questions:

a) How did the presence of soluble salts in the hardcore bring about cracks in the floor?

b) How did the gravel being washed away by the rain bring about the failure of the wall/roof joint?

c) How did the poor quality paint bring about gaps between the window and the frame?

Part 2 Controlled Practices

1. Read this:

Stabilizing the ground under the foundations *prevents* the columns from moving.

Now complete these sentences:

a) Removing the soluble salts from the hardcore _____.

b) Protecting the roof from the sun _____.

c) Painting the woodwork with good quality paint _____.

2. Read this:

We have a problem with the air temperature in this room. It's too cold. This is because of/due to inadequate thermal insulation. You see, to a certain extent, the temperature in the room *depends on* the thickness of the insulation. *Consequently*, we should increase the thickness of the insulation.

Now write similar paragraphs about these:

a) noise level

b) amount of light

c) degree of humidity

3. Read this:

The gravitational force on a structure can be divided into dead loads and live loads.

Dead loads can be calculated accurately because they rarely change with time and are usually fixed in one place. Live loads are always variable and movable, so no exact figures can be calculated for these forces. Structures must also resist other types of forces, such as wind or earthquakes, which are extremely variable. It is impossible to predict accurately the magnitude of all the forces that act on a structure during its life; we can only predict from past experience, the probable magnitude and frequency of the loads.

Engineers never design a structure so that the applied loads exactly equal the strength of the structure. This condition is too dangerous because we can never know the exact value of either the applied loads or the strength of the structure. Therefore, a number called a "factor of safety" is used. The safety factor is denned as the ratio of the probable strength of the structure and the probable loads on the structure. This factor may range from 1 (where there is little uncertainty) to perhaps 5 or 10 (where there is great uncertainty).

Now answer the questions:

a) Can the loads from the internal partitions of a building be estimated accurately? Why?

b) Can the loads from storage in a building be estimated accurately? Why not?

c) How can an engineer predict the possible loads that will occur on a structure?

d) Why do engineers never design a structure so that the applied loads exactly equal the strength of the structure?

e) When there is great uncertainty about the loads on a structure and the strength of a structure, does an engineer choose a high or low safety factor?

f) When does failure occur?

Part 3　Further Development

1. Look and Read:

In some Iraqi houses, a duct is contained between the two skins of a party wall. A wind tower is placed above the duct. This tower faces the prevailing wind with the result that it directs the wind through the duct into the basement of the house. The surfaces of the internal party wall remain at a lower temperature than the rest of the house throughout the day. This is because the wall is very thick and does not receive any direct solar radiation. The incoming air comes into contact with the surfaces of the duct and, as a result, is cooled by conduction. The relative humidity of the air is increased, just before it enters the basement, by passing it over porous water jugs. The air then leaves the basement through an outlet thereby helping to ventilate the courtyard during the daytime.

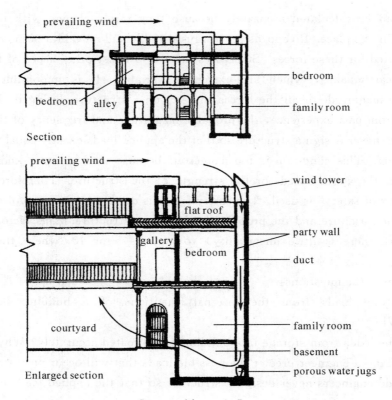

Courtyard house in Iraq

2. Find the relevant cause or causes of these effects and answer these questions:

a) The wind is directed into the basement.

b) The internal party wall is kept cool.

c) The incoming air is cooled.

d) The relative humidity of the air is increased.

e) The courtyard is ventilated during the daytime.

Questions:

a) What material do you think the house is made of? Why?

b) For what reason is the wind tower placed on the roof?

c) Why is the duct placed between the two skins of the party wall?

d) Why are the water jugs porous?

e) Why is the cooled air let into the courtyard from the basement and not from the bedroom?

f) Why is the courtyard in the centre of the house?

g) Where is the entrance to the house?

h) Which bedroom would be best to sleep in?

3. From your own knowledge or from information in previous exercises, decide whether these statements are true(T) or false(F). Correct the false statements.

a) Earthquakes lead to cracking in buildings.

b) If moisture laden air is not allowed to escape from a building, then condensation will occur.

c) Frequent painting of steelwork results in corrosion.

d) Insulating a house allows the air inside to cool to its dew point.

e) The expansion of a roof is caused by a decrease in temperature.

f) In an arch the bricks are wedge-shaped thereby causing their weight to be distributed upwards along the curve of the arch.

g) Houses in hot-dry climates have compact layouts owing to the high solar radiation.

h) A profiled sheet is rigid because of its shape.

i) The contraction of a column is due to tensile forces.

j) Condensation on the surface of a wall causes damp patches and stains and encourages mould growth.

k) Climate affects the form and orientation of buildings as well as the type of materials and construction methods used.

Part 4 Business Activities

Completion Certificates

1. 文化与指南(Culture and Directions)

一个项目或项目的一部分完成后,承包商要递交给业主一份申请书,要求业主颁发工程的或某部分的完工证明书。这种情况下,一个由业主、承包商和政府有关部门联合组成的工程验收小组就成立了。验收小组对合同范围内的所有建筑物进行检查。在验收过程中发现工程中存在的一些缺陷,工程承包商有责任在一定时间内修复不足之处。在完全达到合同要求之后,业主方颁发给承包商完工证明书。通常完工证明书是在一个正式的会议上颁发的。业主可能会宴请(host a banquet)承包商的有关人员,对他们为项目付出的辛勤劳动表示感谢。承包商一方也常致答谢词,感谢业主在项目执行过程中的协助与支持。

2. 情景会话(Situational Conversations)

(As the project is drawing to a close, the contractor submitted his application for issuing the completion certificate of the project. But the owner did not hold the same idea. They are discussing and arguing about it.)

(A: The owner; B: The contractor)

B: Today we would like to submit our application for the section completion certificates for the Turbine Generator Building. The early issuing of the certificate will be very much appreciated.

A: In my opinion, it is too early to issue the section completion certificate for the building.

B: I'm afraid I can not agree with you on this point. In accordance with Clause 32-2 of Conditions of Contract, "A section completion certificate will be issued at the stage building is completed substantially and the function of it can be used by the Employer." We should be given the section certificate since the works has been completed at 95% of the total value of the building.

A: Yes, that's right. But the point is from our calculation you have only completed 92% of the works.

B: I guess your calculation is on the basis of the payment sheet of last month. From then on, 25 days has passed, so the value of the work finished for this building should be already over 95%. Here is the statistics report of the project and the monthly payment application we've prepared for your reference.

A: From my point of view, there are still a lot of items of works unfinished. I would like to point them out one by one if you don't mind.

B: I have no objection to it, if you insist.

A: Let's go over it from bottom to top of the building. First, about 10% of the steel trench covers still remain. Second, there are quite a number of non-slip tiles not yet paved on 16m platform. Third, the final painting to some area of steel structure on 28m has not yet been done. Only these three items are quite enough to cease the issue of the completion certificate for the section works.

B: What you mentioned is true. But you should see the causes of the delay. As we know very clearly that we, as the civil works contractor, are not responsible for such uncompleted works because all the steel trench covers are already fabricated and stored in our warehouse. They can be used to cover the equipment trench as soon as the other contractor has completed the equipment erection. The case is true with your second and third items. I don't think you will agree that the painting works can be done just above heads of erection workers.

A: Apart from what I mentioned, some louvers on wall and several ventilators on roof are not yet handed over to us.

B: The HVAC (Heating, Ventilation and Air-conditioning) works was delayed because of the deferring of your approval of equipment procurement.

A: I am afraid your argument is not reasonable. I noticed the documents left by my predecessor recorded that you proposed a different equipment instead of the one given in specification. It took time to investigate if the proposed equipment was suitable or not.

B: But the investigation period was really unreasonably long.

UNIT 14 POST-CONSTRUCTION PROBLEMS

A: I think we'd better stop arguing here. I suggest that you do your best to finish all the works which are not affected by the erector contractors, such as louvers and ventilators within 30 days. After a further inspection to the whole building, the Certificate of the Completion of the building will be issued to you.

B: Does the retention period start from the day of issuing the Certificate?

A: Sure. But there is one more important thing you should not forget during the period.

B: Yes. I know very well what you are referring to is the as-built drawings. We will have a thorough check of all the as-built drawings and submit them to you within 20 days.

A: That's fine. And other necessary information, such as QA/QC inspection forms, survey records, and trace files should be ready for submission.

B: No problem. But as to the retention money, I wonder whether we can be paid half of the amount, which is 2.5% of the total contract value after we send you a bank guarantee of the same value.

A: I don't think there is a big difference between these two types of the payment. We will think it over and reply to you soon.

(*One year later, Mr. Ordway and Mr. Chen meet again. They are having a table talk.*)

B: We are very pleased to hear that the final completion certificate for whole civil works of the power plant will be issued soon. So we must have a big party for celebration of this most important occasion during these years.

A: I agree with you. It is really worth celebrating. The past three years are very unforgettable, I believe, to both of us.

B: Yes. You are right. The completion of this project means too much for me—it is a real victory in my career.

A: From your words, it seems that you treat the project as a battle and yourself as a general.

B: Yes. For me, it is really a battle. The famous Napoleon Bonaparte said that the secret of winning a war was to mass the effects of overwhelming combat power at the decisive place and time. The same is true with the accomplishment of a construction project. You need to assemble enough material, machine and power at the suitable time and location in accordance with the requirement of schedule and quality.

A: Both organization of battle and construction project need careful planning and good mobilization. But one thing is different. That is in a battle, the enemy is in your opposite side, and he tries to defeat you for his own interests. But in a construction project, you have no opposite side. Apart from your own effort, your client, subcontractors and suppliers help you complete the project.

B: On the one hand, I agree with you that all staff on construction site have a common goal—to complete the works on schedule with good quality and safety. Without fulfilling this common task, nobody can obtain his own benefits. On the other hand, the relationship between the employer and the contractor is complicated, which is half partners and opponents.

A: What you said is a bit fresh to me. Can you explain a bit more on this point?

B: As what you pointed out that the goal of completion of the works is common for both sides, both of us rely on each other. You depend on us implementing the project, from the underground to top level with all structure and decoration works. We depend on your progress payment, drawing supply and so on. We are partners on this point.

A: On the other hand, we are "fighting" each other, are we?

B: Yes, that's true. We work for the benefits of respective companies. Sometimes the benefits are in conflicts. For example, the contractor wants to be paid as soon as possible, and as much as possible, and the vice versa with the employers, at least some employers.

A: What you described is only a partial picture. It is certain that there are different benefits between the employer and the contractor, even though the conflicts are regulated by rules and agreements accepted by both parties. I appreciate the slogan "Fair Play". I think both of us have played fairly in this project and there is no loser here—both parties are winners in this game.

B: You are right. I couldn't agree with you more on this. Now, Let's toast for our good cooperation.

A: And also for our common success and achievement.

A and B: Cheers!

New Words and Expressions：

certificate n.	证书,认证
submit v.	提交,呈送
issue v.	发行,颁布
substantially ad.	大体上,实质上
statistics n.	统计数据
for reference	以供参考
non-slip tiles	防滑砖
fabricate v.	制作,装配
erection v.	安装,建造
louvers n.	百叶窗,(灯笼式的)天窗;气窗
ventilator n.	通风设备,换气扇

procurement	n.	采购,取得
defer		推迟,延期
predecessor	n.	前任,前身
specification	n.	规格,规范
refer to		涉及,指的是
as-built drawings		竣工图
trace files		跟踪文件
retention money		工程累积保证金
mobilization	n.	动员,调动
subcontractor	n.	分包商
opponent	n.	对手,敌手
progress payment		工程进度款
respective	a.	分别的,各自的
conflict	n.	冲突,矛盾
partial	a.	部分的
fair play		公平竞争

Notes to the Text:

1. I guess your calculation is on the basis of the payment sheet of last month. 我想你的统计是根据上个月底的付款申请书的数字。

2. I have no objection to it, if you insist. 如果你想这么做我也不反对。

3. The same is true with the accomplishment of a construction project. 成功地完成一个工程项目也是同样道理。"The same is true with sth." 是一个固定表达,表示"……同样如此""……也适用于此"。

4. The famous Napoleon Bonaparte said that secret of winning a war was to mass the effects of overwhelming combat power at the decisive place and time. 拿破仑·波拿巴曾经说过,他在战场上获胜的秘诀就是集中兵力,在决定性的地点与时间集中作战力量。

5. What you said is a bit fresh to me. 你的这番话还是挺有创见的。这里的 fresh 表示(谈话等)"有独创性的;有启发性的,生动的"。

6. For example, the contractor wants to be paid as soon as possible, and as much as possible, and the vice versa with the employers, at least some employers. 其中"the vice versa"意思是"反之亦然",这句话的意思是"作为承包商,他们希望能尽可能快、尽可能多地收到工程款,而业主(至少说有些业主)的想法就恰恰相反(他们想尽可能晚、尽可能少地支付工程款)。"

7. I couldn't agree with you more on this. 在这一点上我非常赞成你的看法。

Post-Reading Exercises:

Complete the sentences with the proper words or expressions given in the box.

| specification | ventilator | erection | procurement | louvers |
| as-built drawings | retention money | progress payment | | |

1. _____ is one of a set of parallel slats in a door or window to admit air and reject rain.

2. _____ is a device (such as a fan) that introduces fresh air or expels foul air.

3. This sum is know as _____ and serves to insure the employer against any defects that many arise in the work

4. Contractor shall produce _____ for single line diagrams, panel schedules, schematic and wiring diagrams showing any changes made during performance of the work.

5. Legislation will require UK petrol companies to meet an EU _____ for petrol.

6. The _____ is an installment of a larger payment made to a contractor for work carried out up to a specified stage of the job.

7. All site _____ works will be performed by the buyer under the technical instruction of the seller.

8. Change is clearly needed, because government _____ practices have turned out to be harmful.

Part 5 Extensive Reading

(A) Fatigue

Fatigue failure occurs in members subjected to variable loads at values significantly below those that would cause failure under static conditions. In static tests to failure, there is a large elongation prior to failure, whereas in fatigue failures there is no plastic elongation.

Structures subjected to fatigue loading include bridges, crane girders, conveyer gantries, mine head-frames, offshore oil rig platforms, etc. Normally the fatigue crack forms at the surface and progresses into the body of the material growing normal to the direction of the cyclic tensile stress body of the material by repeated opening and closing of the crack tip. In welded structures the crack may start at a weld defect. The fatigue crack has a smooth surface with markings emanating from the start of the crack. Failure occurs when the crack has grown large enough so that the remaining section cannot carry the load.

The usual laboratory test is the rotating bending test made on small cylindrical specimens with or without notches which are loaded as cantilevers or beams. As the specimen rotates, the stress at any point varies between equal maximum values of compression and tension. For a given value of maximum stress S, the specimen is tested until failure occurs and the number of cycles N at failure is noted. The test is repeated for other values of stress. The relationship between the applied stress and number of cycles to cause failure is found. The fatigue endurance limit or stress below which the specimen can sustain an infinite number of cycles can be found.

The above is not a great deal of use in the design of welded structural steelwork where stress concentrations and defects in the welds initiate failures. The results of tests on welded joints must be used as a guide. Stresses causing failure fob a given number of cycles, i. e., life, are determined for particular welded connections.

Structures are usually subjected to widely varying or random stress cycles. Methods are available for estimating the cumulative damage caused by a given load spectrum. The method is based on the S-N curve for the critical weld detail to be used and the limiting stress to give the required number of cycles or life is determined.

Notes to the Text:

1. In static tests to failure, there is a large elongation prior to failure, whereas in fatigue failures, there is no plastic elongation. 在对损坏的静态测试中,损坏之前存在着一个较大的延伸率,然而,在疲劳损坏里是不存在塑性延伸率的。

2. As the specimen rotates, the stress at any point varies between equal maximum values of compression and tension. 当样品转动时,任何一点上的应力在压力和拉力相同的最大值之间变动。

3. Stresses causing failure for a given number of cycles, i. e. life, are determined for particular welded connections. 对于特定的焊接,产生损坏的给定循环次数(即寿命)的应力是确定的。

4. Structures are usually subjected to widely varying or random stress cycles. 结构通常承受变化非常大的或随机的应力。

5. The method is based on the S-N curve for the critical weld detail to be used and the limiting stress to give the required number of cycles or life is determined. 这种方法依据用于关键焊接点的 *S-N* 曲线。该曲线上的循环次数(或寿命)的极限应力是确定的。

New Words and Expressions:

fatigue	*n. v.*	疲劳,疲化
static	*a.*	静态的
elongation	*n.*	延长,伸长,延伸率
prior to		在……之前,先于
crane	*n.*	起重机
girder	*n.*	横梁,吊臂
conveyer	*n.*	输送机,传送机
gantry	*n.*	吊架,桥形台架

mine head frame		矿井架
emanate	v.	流出,生出
notch	n.	(V形)切痕
cantilever	n.	悬臂
grind	v.	摩擦,转动
flush	n. v.	冲洗,流动
flange and web		凸缘与腹板
stiffener	n.	加劲杆(条);支肋
splice	n.	接头,接合处
intermittent	a.	间歇的,中断的
compression	n.	压力
spectrum	n.	光谱,频谱
random	a.	随意的,随机的
available	a.	有效的,有用的

Post-Reading Exercises:

Choose the one that best completes the sentence.

1) How does the fatigue crack normally happen in a certain material?

2) What is the main idea of Paragraph 3? _____

  A. The stress changes in the range between maximum values of compression and tension.

  B. In the test to fatigue, the relationship between stress and number of cycles is found.

  C. The fatigue endurance limit or stress can be found in the laboratory test.

  D. When the specimen rotates, the related C-N curve is changed.

3) What affects the performance of welds? _____

  A. Stress concentrations   B. Abrupt changes of section

  C. Weld defects   D. all of above

4) The best results of the test to girders fabricated with fillet weld between flanges and webs are got from _____.

  A. The continuous and automatic process weld

  B. the intermittent weld

  C. low fatigue strength

  D. high fatigue strength

5) By what means can the cumulative damage caused by a given load spectrum be estimated. _____

  A. The limiting stress.

  B. The number of cycles.

  C. The skill of a w elder.

  D. The S-N curve for the critical weld detail.

(B) Bridge Deficiencies

Bridge deficiencies evolve from a variety of situations and conditions. Basic design criteria, traffic usage, environmental factors and other site conditions are all involved to some extent and are responsible for specific deficiencies. An additional, and perhaps the most important, contributor to bridge deficiencies is the level of maintenance employed. Deficiencies' causes can be categorized into two main areas: (1) inherent deficiencies, which result from the design of the facility and (2) those deficiencies, which result from the use of the facility. Deficiencies from either cause are subdivided into four areas: structural, mechanical, safety and geometric deficiencies.

1. Structural deficiencies

Structural deficiencies are caused most frequently by lack of proper maintenance, poor design details and light original designs. In steel structures, paint system breakdown permits corrosion of the base metal to begin. Once started the process accelerates as large areas become exposed. Eventually the metal corrosion can result in section loss serious enough to have an impact on the load-carrying capacity of the member. If left uncorrected, the process will continue, resulting in the ultimate collapse of the bridge. Concrete members also deteriorate at a rapid rate when exposed to adverse environmental conditions. Penetration of brine solution through the unprotected concrete surface causes the reinforcing steel to oxidize and expend, ultimately leading to cracking of the concrete cover.

Once the process begins, it accelerates at a rapid pace as more of the corrosive materials reach the reinforcing steel.

Many of the structures were not designed initially to carry the loads being imposed on them by modem traffic. These deficiencies apply to sub-structure members as well as to superstructure members. In addition, substructure elements can be structurally deficient because of foundation conditions.

2. Mechanical deficiencies

Mechanical deficiencies are primarily caused by corrosion of metal elements, the accumulation of debris and silt around bearings and joints and poor design details. For example, built-up of debris around bearing areas often completely covers metal bearings. This debris is composed of bird droppings, nesting materials and other deposits that are highly corrosive. The bearings freeze as a result of this corrosion and prevent the bridge from functioning as intended.

Settlement as well as lateral movements in piers can cause a similar situation. Pier rotation causes roadway joints to close and bearings to exhaust their capability to accommodate movement. Temperature changes can then result in serious overstress in other elements of the bridge.

3. Safety deficiencies

Many safety deficiencies are geometric in nature, e. g. roadway width Some safety deficiencies occur because structure members are located in a position, where they become a hazard to the motorist, such as end posts or pylons placed close to lanes.

New Words and Expressions:

evolve from		从……中进化(演化)
criteria	*n.*	标准
maintenance	*n.*	保持；维护；生活费用；抚养
inherent	*a.*	固有的；生来的；内在的
original	*n.*	最初的；新颖的；原物，原文
corrosion	*n.*	腐蚀；侵蚀；锈
capacity	*n.*	容量；能量
reinforce	*v. n.*	加强；加固物
oxidize	*v.*	(使)氧化；(使)生锈
utilize	*v.*	运用，使……有用，运用
vehicle	*n.*	车辆
superstructure	*n.*	上部构造，上层建筑
accumulation	*n.*	积聚；堆积物
rotation	*n.*	旋转；自转；循环；轮作
pier	*n.*	(桥)墩；码头；支柱
geometric	*a.*	几何图形的，几何学的
abutment	*n.*	(建)桥台；拱座
pylon	*n.*	塔门；路标塔；架高压电线的铁塔

Post-Reading Exercises:

1. Choose the one that best completes the sentence.

1) Basic design criteria, traffic usage, environmental factors and other site conditions _____ specific deficiencies.

  A. lead to   B. lean on   C. are restricted to   D. rely on

2) Eventually the metal corrosion can result in section loss serious enough to _____ the load-carrying capacity of the member.

  A. support   B. link

  C. connect   D. influence

3) Penetration of brine solution through the unprotected concrete causes the reinforcing steel to oxidize and expend, ultimately leading to surface serious _____ of the concrete cover.

  A. damage   B. danger

  C. impact designed   D. influence

UNIT 14 POST-CONSTRUCTION PROBLEMS

4) Many of the structures were not designed _____ to carry the loads being imposed on them by modem traffic.
A. from the beginning   B. exclusively
C. inclusively   D. from the bottom

5) Temperature changes are _____ for other elements of the bridge
A. dangerous   B. available   C. important   D. essential

6) The bearings freeze as a result of this corrosion and prevent the bridge from functioning as _____.
A. said   B. wanted   C. promised   D. indicated

7) The debris is made of bird droppings, nesting materials and other deposits that are _____ corrosive.
A. absolutely   B. to some extend   C. very   D. deeply

8) Some safety deficiencies occur because structure members are located in a position where they become a hazard to the _____.
A. passenger   B. customer   C. driver   D. car

2. Decide whether the following statements are true(T) or false(F) according to the passage given above.

1) Deficiencies due to the use of the facility are called inherent deficiencies. ( )

2) Structural, mechanical, safety and geometric reasons may lead to various deficiencies. ( )

3) In steel structures, a possible reason for corrosion is a breakdown of the paint system. ( )

4) Once the process of oxidation begins, it will accelerate at a slow pace, even if more of the corrosive materials reach the reinforcing steel. ( )

5) Superstructure elements can be deficient because of poor foundation. ( )

6) Structural deficiencies are mainly caused by corrosion of metal elements, the accumulation of debris and silt around bearings and joints and poor design details. ( )

7) Some animals—such as birds, can cause massive damage to steel bridges and lead to heavy corrosion. ( )

8) Many safety deficiencies are caused by geometric reasons. ( )

Keys to the Exercises from Unit 1 to Unit 14

Unit 1

Part 1　Warm-up Activities

1. 立方体；半球体；三棱柱；金字塔形；长方体；圆锥体；圆柱体

3.
a) a minaret
b) a beam
c) a steel channel
d) an Egyptian house
e) an Arabic arch
f) an church

Part 2　Controlled Practices

1.
a) pyramid
b) is shaped like a cone.
c) is shaped like a hemisphere.
d) of the structure of the building is shaped like a cylinder.
e) of the structure of the building is shaped like a rectangular prism.
f) is shaped like a triangular prism.
g) is shaped like a cube.

2.
The cross-section of the brick is rectangular in shape.
The cross-section of the hotel is square in shape.
The cross-section of the top of the minaret is circular in shape.
The cross-section of the column is circular in shape.
The cross-section of the church is rectangular in shape.
The longitudinal-section of the brick is rectangular in shape.
The longitudinal-section of the hotel is triangular in shape.
The longitudinal-section of the top of the minaret is triangular in shape.
The longitudinal-section of the column is rectangular in shape.
The longitudinal-section of the church is triangular in shape.

3.

a) The church is hollow. It has four flat external surfaces.

b) The slab is solid. It has six flat external surfaces.

c) The column is solid. It has a curved external surface.

d) The mosque is hollow. It has a curved external surface.

e) The steel beam is solid. It has eight flat external surfaces.

Part 3 Further Development

1) 规划、设计、施工和管理/灌溉和排水系统到火箭发射设施

2) 道路、桥梁、隧道、大坝、港口、发电站、水系统和污水系统、医院、学校、公共交通系统

3) 私人拥有的/机场、铁路、管线、高楼大厦和为工业、商业、民用

4) 规划、设计和修建整个城市和乡镇/空间站

5) 工程的类型/各种技能

6) 土木工程师/岩土工程专家

7) 环境专家/潜在的空气污染和地下水污染/满足政府对……要求

8) 运输专家/什么类型的设施/对当地道路和其他运输网

9) 施工管理专家/其他专家/订购工作所需的材料和设备/做其他的监督管理工作

10) 任何给定的工程/广泛地利用计算机/被用来设计工程的各个部分

Part 4 Business Activities

1. Tick the statements you think are true.

1) T 2) F 3) T 4) T 5) F 6) T 7) F 8) T 9) T 10) F

2. Translate the following sentences into Chinese:

1) 基于安全理由,(飞机)起飞与降落时期禁止使用电子设备,如手机、电脑、CD 播放机、MP3/MP4 和游戏机。

2) 请您务必将行李放在行李架上或坐椅下方。

3) 客舱出口、过道以及紧急出口处不得摆放任何行李。

4) 请注意,30 分钟后,我们将在洛杉矶机场降落。为了您的安全,请务必系好安全带。

5) 请带好您的护照、机票、相关的入境申报表及全部的私人物品到候机厅办理出入境手续。

Part 5 Extensive Reading

(A)

1.

1) 独门独户房

2) 作为遮风挡雨的设施

3) 按布局或平面图可划分为……

4) 陡坡形的屋顶

5) 半组楼梯

6) 卧室的间数

2. 1) T 2) F 3) F 4) F 5) T 6) T 7) F 8) F

3.

1) A house is a building that provides home for one or more families. Its main function is to provide shelter from element, but it usually serves many more purposes. It is a center of family activities, a place for entertaining friends, and a source of pride in its comfort and appearance.

2) Builders, rather than professional architect, can design buildings in the US.

3) Cape Cod; Two-Story House; Ranch House; Split-level House; Attached Houses; Row Houses.

4) Cape Cod

5) Ranch House

6) Split-level House

7) Cape Cod

8) In the mid-1960s, a most important value in housing was sufficient space both inside and outside, many families preferred to live far away from the center of a metropolitan area as possible to get away from noise, crowding and the confusion of the city center. However, people today require much more than this of their housing. In regard to safety, health and comfort, people's requirements have changed.

(B)

1.

1) 从事建筑设计这个职业

2) 与时俱进

3) 在大学获得学位

4) 在大学获得建筑设计硕士学位

5) 参加建筑设计比赛

2.

1) T 2) T 3) F 4) F 5) T 6) F 7) F 8) T

3.

1) Be curious about the surroundings and interested in learning how to improve them etc.

2) He/she should have the imagination to create the buildings and cities our society needs to keep pace with the changing world.

3) Economics.

4) Manchester University, University of Westminster and South Bank University.

5) No.

6) Architectural assistant, editor, acting as a judge for architectural competitions.

Unit 2

Part 1　Warm-up Activities

1. 天然土壤；耐火砖；网状材料；纤维材料；夯实土壤；空心砖；液体；松散材料；沙子，

灰和土；混凝土；玻璃；金属；沙土；钢筋混凝土；橡胶；天然石材；矿渣（粉），塑料；木材；碎石；多空（透气）材料；防水材料；胶合板；普通砖；釉面砖；抹灰层；石膏板。

2. （略）

3.

Steel has the property of high tensile strength. This means it can resist high tensile forces.

Stone has the property of high compressive strength. This means it can resist high compressive forces.

Glass wool has the property of good thermal insulation. This means it does not transmit heat easily.

Brick has the property of good sound insulation. This means it does not transmit sound easily.

4.

Because glass is transparent；

Because glass wool has the property of good thermal insulation；

Because zinc is corrosion resistant；

Because asbestos is non-combustible；

Because the corrugated shape makes the sheet resist high tensile forces；

Because concrete has high compressive strength；

Part 2 Controlled Practices

1. A __4__ B __6__ C __2__ D __3__ E __1__ F __7__ G __8__ H __5__

2. 1) impermeable 2) corrosion resistant 3) heavy 4) a good conductor of heat 5) not a good conductor of heat 6) rigid 7) non-combustible 8) opaque

3.

Material	Availability	Use	Property	Problems/Durability
cane leaves vine bamboo palm-fronds	warm-humid zones	glued board, furniture, etc.	combustible/light	combustible
grass	mild zones	glued board, furniture, etc.	combustible/light	combustible
hardwood softwood	all cold, mild and humid zones	glued board, furniture, etc.	combustible/light	combustible
earth	continental area	brick, tiles, etc.	non-combustible	permeable
concrete	everywhere	structure, floor, etc.	non-combustible	impermeable/heavy/durable,

4.

Heavy form of material	Function of components		
	Structural support only	Space dividing only	Both structural and space dividing support
block	√	√	√
sheets		√	
rods		√	

5.

a) F   Rod material can only be used for dividing space.

b) F   Concrete can only be used as a block material and a rod material.

c) F   Steel is used for frame construction because it has high tensile strength and high compressive strength.

d) T

e) F   Mass construction buildings are heavy whereas planar construction buildings are light.

6.

A copper tube is an example of a rod, because it is light.

A concrete block is an example of a compressive material, because it is heavy and rigid.

A steel stanchion is an example of a block, because it has high tensile and high compressive strength.

Part 3   Further Development

1.

1) What is the feature about?

Your sentences: What is the feature about this new construction material?

2) answer our purpose

Your sentences: This kind of gasoline answers our purpose satisfactorily.

3) The average... of ... is

Your sentences: The average age of the boys in this class is fifteen.

4) come from

Your sentences: Illness may come from a poor diet.

5) compare with

Your sentences: My English cannot compare with hers.

Part 4   Business Activities

1) B: We could provide relevant data to you, but you are responsible for your own interpretation of them.

2) B: We don't think that we are responsible for the damage. As stipulated in the contract, once a part of the works has been handed over to the employer, the contractor shall not have the obligation to protect it. Further, the damage is obviously due to vandalism.

3) A: If we accept the subcontractor nominated by you, do we have to hold ourselves responsible to you for his or his employees' behavior?

B: Yes, as the general contractor, you will be responsible for all the consequences arising from their misconduct.

Part 5　Extensive Reading

(A)

1.

1) Masonry, timber, steel, aluminum, concrete, plastics.

2) Two types of behavior can occur: brittle and plastic.

3) No, only softwoods suit.

4) Aluminum.

2.

1) elasticity, stiffness　2) elastically　3) Timber　4) Masonry

5) Steel　6) Aluminum　7) Portland cement　8) original

(B)

1. 1) T　2) F　3) F　4) T　5) T　6) T　7) F　8) T

2.

1) Since wood is one of the few materials with good tensile and insulation properties, it is mainly used in framing, for example, to make doors, window frames and furniture.

2) Laminated wood is made of many small strips of wood that are glued together or joined together with mechanical fastenings to form a large piece of wood.

3) Laminated wood is available as big structural elements in any desired size including curves and angles.

4) Wood panel products are big timber elements made of several thin layers of veneers that are glued together.

5) There are 3 ways of slicing the veneer: rotary slicing, plain slicing and quarter slicing. For rotary slicing, the timber logs are soaked in hot water to soften the wood, then each is rotated in a large lathe against a stationary knife that peels off a continuous strip of veneer, much as paper is unwound from a roll. Plain slicing is to slice plainly along the length of the timber; quarter slicing is to slice quarterly along the length of the timber.

6) The timber logs are soaked in hot water to soften the wood.

7) The sheets are repaired as necessary with patches glued into to fill open defects.

8) The most economical method is rotary slicing, which is used for veneers for structural wood panels. The finest figures are produced by quarter slicing, which results in a very close grain pattern with prominent rays. The grain figure produced by rotary slicing is extremely broad and uneven. For better control of the grain figure in face veneers, the veneers are plain sliced or quarter sliced.

Unit 3

Part 1 Warm-up Activities

2. a) dining area   b) living room   c) kitchen   d) hall   e) toilet   f) bedroom   g) bathroom   h) bedroom   i) terrace

Part 2 Controlled Practices

2.

a) T

b) F   There are 3 adjacent bedrooms on the first floor.

c) T

d) F   There is a cupboard between the kitchen and the dining room.

e) T

f) F   There is a hall under the hall.

g) F   Bedrooms occupy most of the first floor.

h) T

i) T

j) F   From the garage, you pass through the kitchen to enter the family room.

k) F   The entrance is situated in front of the stairs.

l) T

m) T

3.

1) receive a commission for a building

2) submit preliminary plans and a rough estimate of the cost to the client for his approval.

3) incorporate into

4) bid for

5) make periodic inspection to

6) defects liability period

7) takes full possession of the building

4. a) design   b) commission   c) visit site   d) preliminary plans and rough estimate of cost   e) plans   f) incorporate   g) tender   h) draw up a contract   i) client   j) contractor   k) site   l) bills   m) bills   n) payments   o) completes   p) occupies   q) defects liability period   r) take full possession of the building

Part 4 Business Activities

Post-Reading Exercises：

1)项目管理部门下设五个部门：工程部、技术部、物资部、行政部、质量（监控）部。

2)现在我们来讨论第五章：施工机械计划表。我们很欣赏你们的施工机械计划，为本工程调运了必需的设备，不过计划里仍有设备漏项或不足。

3)你无法保证高峰时段，在没有备用机的情况下这些设备都能正常运转，而且你的设备看上去都不够达标，它们占用的空间太大，而且材料的计量也不精确。因为这是关键设备，我方建议贵方使用新的先进的搅拌机型。

Part 5　Extensive Reading

(A)

1.

1) C　2) D　3) B　4) A　5) A　6) B　7) A　8) B　9) A　10) C

2.

1) When man for the most part abandoned cave dwelling, cave-like shelters appeared. They were simple in design and their construction matched the technology of that day. Drawings were not necessary.

2) In the last few hundred years, architectural drawing has evolved into several general types. These range from concept sketches to intricate details drawn to scale. Usually, clients are not trained in grasping concepts from rough sketches. Therefore, a more picture-like drawing, known as a presentation drawing, is required to explain a proposed building to the prospective owner. Presentation drawings are usually necessary for commercial work and custom homes, while homes built on a speculative basis with no owner committed usually do not require this preparation. Once the design has been accepted, drawings must be prepared to guide the builders of the project. These are called working drawings. They are precisely drawn and include plan views, elevations and details, all with dimensions and notes.

Another type of architectural drawing is one that is completed after the construction of a building. It is called an as-built drawing if it contains dimensions and other technical data. Architectural drawing is the means of graphic communication used within the professions and businesses that are concerned with the design and construction of buildings.

3) Design sketches are rough drawings that are used as "idea sketches", made to explore concepts that will be refined at a later date. They may appear crude to the casual observer, but a closer study of sketches drawn by talented designers will usually reveal a theme and sensitivity containing the elements of good design. It is the purpose of these drawings to establish such elements. Design sketches have changed very little from the earliest known examples to those of today's architects.

4) Often, modern design presentation drawings are characterized by realistic features, such as shades, shadows, people, trees, plantings and automobiles.

5) It may be for use in further technical work, for maintenance of the building, or it may be used for publication.

6) The building profession falls generally into two categories, residential and nonresidential. Nonresidential includes commercial, institutional industrial, recreational and other types of buildings that are not houses. Residential work includes multi-family apartments, condominiums town houses and single family buildings.

7) Although state laws vary, typically it is a legal requirement that new nonresidential and multifamily buildings be designed by registered architects.

8) To become a licensed architect, one typically earns a degree in architecture from an accredited university, serves an apprenticeship under a registered architect, and then passes a lengthy examination.

(B)

1.

1) D 2) D 3) B 4) B 5) C 6) C 7) B 8) A

2.

1) F 2) F 3) T 4) F 5) T 6) F 7) T 8) T

Unit 4

Part 1 Warm-up Activities

4.

a) roof beam and floor beam	b) steel stanchion
c) steel stanchion, roof beam and floor beam	d) wedge shaped blocks
e) 4m	f) tie(beam)
g) 3m	h) 1m
i) wedge	j) stone, steel, brick, concrete, etc.

Part 2 Controlled Practices

1.

1) A roof looks like a slab.

2) A roof consists of roof structure and water proof.

3) The floor structure is made of vinyl tiles and concrete panels.

4) Corrugated sheets are made of steel.

2.

a) timber	b) joist
c) asphalt	d) vinyl tiles
e) concrete panel	f) corrugated sheets
g) steel stanchion	h) concrete

3.

The factory is made from four elements: the roof, the walls, the floors and the foundations. The roof has a waterproof covering, which is made of asphalt, and a roof structure, which is made of timber joists and wood-wool slabs. The walls are constructed from two elements: the wall structure, which consists of structure and the cladding wall, which is made of corrugated steel sheets. The surface floor consists of a wearing surface, which is made of vinyl tiles and a floor structure, which is made of precast concrete panels. The foundations consist of concrete column bases and precast concrete piles.

4.

a) To make the frame structure stronger.

b) Asphalt is a very good waterproofing material.

c) Steel sheets can be easily connected with the steel stanchions.

d) Vinyl is used for the wearing surface because it is lighter than the cement floor and ceramic tiles floors.

e) Concrete is used for the column bases to reinforce the foundation.

Part 3　Further Development

2.

a) Steel angles are fixed across the ends of the beams and built into the brick walls. These angles tie the frames together and also provide a place to fix the top of the cladding.

b) The beams are bolted to steel stanchion caps. The stanchion caps are welded to the top of each stanchion. The load on each beam is transmitted through these plates to the stanchions. The bottom of each stanchion is welded to a base plate. Each base plate is fixed to a concrete column base by two holding-down bolts. In this way, the loads of a roof beam are transmitted to the column base.

c) Mortar bed.

d) The roof beams cantilever a short distance beyond the stanchion so that they can extend over the profiled sheet steel cladding, which can be placed outside the line of the stanchions.

Part 4　Business Activities

Post-Reading Exercises：

1) 首先我要阐明的是，该合同条款符合FIDIC《土木工程施工合同条件》的基本原则。

2) 我方建议增加一项附加条款，如：一旦承包商遭遇图纸供应不及时或延误，他有权利获得合同工期的延期许可，同时弥补工期所需的费用也必须追加到合同总价中。

3) 对于如此重要的事项有必要增加一项附加条款：当承包商认为由于不能按时得到图纸(的原因)将要导致延误工期，承包商必须在合理的时间内给工程师(业主方)呈送书面通知书。

Part 5　Extensive Reading

(A)

1.

1) 承受结构的重量

2) 由建筑师设计

3) 通过图解的方法对这个结构进行分析

4) 通过结构的荷载

5) 典型的多层结构

6) 实心地板和空心肋形地板

7) 浇注具有贯穿式孔洞的肋形楼板

8) 悬挂结构

9)减少曲压效应
10)承载电梯、楼梯、风道及电缆
11) the design of building structures
12) office buildings used as dwellings
13) two buildings built in London
14) a hollow concrete tower some 12m square
15) the core of the building
16) high-tensile steel bars dropping from a bridge
17) to hide in a door frame or window frame
18) no noticeable obstruction to sight
19) the growing of the idea of design
20) joined by lightweight bridge

2.
1) suspended structures   6) the core of the building
2) the columns   7) a bridge
3) the bulking effects   8) steel bars
4) one column   9) thin
5) a hollow concrete tower   10) hidden

3.
1) A structure is the part of a building that carries its weight.

2) To make out how the loads pass through the structure with the particular members chosen.

3) The most important member which the engineer designs is the floor.

4) Ribbed floors are lighter than solid floors.

5) The columns of suspended structures are fewer and larger so as to reduce the bulking effects on them and to increase their effective length.

(B)
1. 1)C 2)B 3)D 4)B 5)B 6)D 7)D 8) C
2.
1) Beams, arches truss and columns

2) The beam is called a simply supported or simple beam. It has supports near its ends, which restrain it, only against vertical movement. The ends of the beam are free to rotate. The beam is a cantilever. It has only one support, which restrains it from rotating or moving horizontally or vertically at that end. When a beam extends over several supports, it is called a continuous beam.

3) A cantilever is a long piece of metal or wood used in a structure such as a bridge. One end is fastened to something and the other end is used to support part of the structure.

4) An arch is a curved beam. It differs from a straight beam in the following ways:

(a) Loads induce both bending and direct compressive stresses in an arch. (b) Arch reactions have horizontal components even though loads are all vertical. (c) Deflections have horizontal as well as vertical components.

5) The column is an essentially compression member. The manner in which a column tends to fail and the amount of the load causes failure depends on the material of which the column is made, and the slenderness of the column. The shape of a column is also very important. For example, a sheet of cardboard has practically no as strength as a column, but if bent to form an angle section or other shapes, it is capable of supporting a load.

6) A truss is a framed structure consisting of a group of triangles arranged in a single plane in such a manner that loads applied at the points of intersection of the members will cause only direct stresses(tension or compression) in the members.

7) Yes, it can.

8) Direct stresses.

Unit 5

Part 1 Warm-up Activities

1.
a) 条形基础
b) 平板基础
c) 台柱下条形基础
d) 墙下条形基础

2.
Concrete Foundation
1) 竖墙
2) 起始筋,搭接钢筋
3) 支撑模板
4) 水平施工缝
5) 天然土壤
6) 现场混凝土
7) 垫层
8) 承载分布的角度
9) 天然斜坡的角度

Reinforced Concrete Foundation
1) 竖立柱
2) 钢筋
3) 箍筋
4) 施工缝
5) 折弯的钢筋
6) 垫层

7) 底层土

8) 低承载性地层

9) 直形钢筋

Part 2 Controlled Practices

2.

1) Dead load is a fixed position gravity service load, so called because it acts continuously toward the earth when the structure is in service.

2) Permanent ones and transient ones.

3) Since snow has a variable specific gravity even if one knows the depth of snow for which design is to be made, the load per unit area of a roof is at best only a guess.

4) All structures are subject to wind load, but they are usually only those more than three or four stores high, other than long bridges for which special consideration of wind is required.

3.

令人满意的建筑物的基础必须满足三个基本要求：(1)基础，包括底层的土壤和岩石，必须是安全的，不能产生导致结构崩溃的结果。(2)在建筑物的寿命期限内，基础的沉降必须不能损坏到结构或损害其功能。(3)基础的建造无论从技术上和经济必须是可行的，而且对周围的建筑无不良影响。

Part 3

2.

1) The purpose of a foundation is to carry the load of a structure and spread it over a greater area. 2) The bearing capacity of a soil means the maximum load per unit area on which the ground can safely support the structure;

3) As the nature of the soil often varies considerably, on the same construction site, the capacity of the soil, which supports the loads also varies considerably.

4) It is not always possible to provide a uniform size of foundation for the entire structure.

5) a foundation normally consists of either plain or reinforced concrete, which should lay sufficiently below the ground frost level to avoid the possible danger of frozen soil which will lift it.

6) Soft sports are usually filled with consolidated hardcore or a weak concrete.

7) Isolated columns or stanchions are normally supported on square concrete foundation bases.

8) Such columns are spaced at close intervals.

9) It is often more practical to provide a continuous concrete strip foundation to carry a complete row.

10) A raft foundation is often recommended to support normal buildings.

3.

1) 基础的大小和类别取决于土壤的性质和所负荷的承载。

2) 在同一处建筑工地，土壤的性质不同，其承载力也不同。

3)在用小间距的柱子时,最好的办法是采用连续的条形基础以承载一个整排的重量,就如给一堵承重墙做基础一样。

4)在承载力较差或者土壤质量差异较大的情况下,建议采用筏形基础以承载一般建筑。

5)当基础土壤较差,其开挖深度应该比正常深度大,直到接触到坚实土壤为止。

Part 4 Business Activities

1)本工程的标价是根据我们修建电厂的经验作出的。我们已将利润降低到很低的水平了。考虑到业主的利益,我们相信工程以优良的质量和合理的价格竣工要比价格低而进度质量存在问题这种情况要好得多,因为施工费用与提前发电产生的效益相比要小得多。

2)这儿有几种我们常用的机械折旧年限。例如,大型耐用的设备为15年,一般设备为10年,而易消耗的机械为3年。对这样工期紧张的大型工程,必须采用施工设备清单中许多高效率的设备如混凝土泵、大容量的混凝土搅拌站、混凝土搅拌车和许多种类的吊车等等。

3)第5项只是当地人工的费用。我们将培训当地工人对如此规模的工程怎样进行安排、怎样计划施工程序、怎样操作施工设备和测试仪器。第6项是为了一个完备的质量保证措施的运行而使所有的人员采取必要的措施的费用。例如,对材料供应商的调研、材料的测试将比通常进行得更仔细。

Part 5 Extensive Reading

(A)

1.
1) D 2) C 3) C 4) B 5) A 6) B 7) B 8) A 9) C 10) A

2.

1) During the design development phase you should be considering the various types of structural systems that might be used for your building. The system chosen will undoubtedly affect the layout and appearance of the building.

2) The continuous spread footings spread the weight of a building over a broad enough area so that the soil is not penetrated by the foundation, allowing to move downward. If some areas of the soil are softer than others, the foundation should still span over these areas without sinking.

3) Soil conditions, economics, and occasionally esthetic considerations.

4) Lateral forces

5) 14 inches

6) A grid pattern of piling will result in a network of sound supports that can be spanned by beams to support the building weight over the undesirable soil.

7) The expanding earth is strong enough to lift entire buildings. This lifting and lowering(upon melting of the frozen moisture) of buildings or their parts causes cracking of walls and floors because the movement is rarely evenly distributed.

8) Wood, steel and round concrete piers are essentially short piling: They are appropriate when great depth is not needed to achieve sound support. If a spread concrete

footing(pad) is used, masonry may be added to the list of pier materials.

(B)

1. 1) A 2) A 3) B 4) A 5) B 6) D 7) A 8) A
2. 1) F 2) F 3) F 4) T 5) F 6) T 7) F 8) F

Unit 6

Part 1 Warm-up Activities

2.

a) The external walls are made up of brick cladding, wall planks, windows, doors, heads and sills, stanchion casings and inner lining panels.

b) While the steel frame is being erected, the wall planks and floor units are fixed.

c) At the same time, the stanchions are enclosed in casings, which serve the function of resisting fire.

d) The precast concrete floor units are capable of carrying a load of up to 5kn/sqm.

e) The wall planks are designed to be weatherproof and to support the outer cladding.

f) The 900mm and 1,800mm wide external doors are installed.

g) The glazing is done on site.

h) The internal sills and lining panels are installed to form a cavity for the heating and electrical services.

i) A grill is underneath the sill.

j) A grill underneath the sill, together with an air intake at skirting level, enables air to circulate up past the finned heating element.

k) The lining panels are capable of being removed to give access to the services.

3.

a) It's much safer to install them from inside building for security purpose.

b) to spread the heat better.

c) No, they are pre-glazed.

d) Aluminum is much lighter and anti-corrosive than steel.

e) Terrazzo and granite.

f) School, hospital or residential.

Part 2 Controlled Practices

1.

a) 2∶1

b) better

c) direct proportional

d) compare, brick wall

e) thicker

Keys to the Exercises from Unit 1 to Unit 14

Part 4 Business Activities

Post-Reading Exercises：

1. "… happens to do (be) sth." is a very useful expression, meaning "正巧……"。

1) During the excavation, some antiques happened to be discovered.

2) B：I happened to see him when I was doing shopping in the supermarket yesterday.

3) B：Fortunately, all the people happened to be out of the tunnel when the cave-in occurred. So no one got injured.

2. "according to" is used quite often during a negotiation, meaning "根据,按照"。

1) B：Because according to the weatherman, there will be a heavy rain tomorrow.

2) You should have issued to us the taking-over certificate for the section by the end of last month according to Clause 15 of the contract.

3) B：According the technical specifications, all the cement shall be delivered to the site in sealed bags of standard weight. Bulk cement is not acceptable.

3. "reserve the right to" "保留……权利"。

1) This event is force majesty, we reserve the right to lodge a claim for it.

2) As this is just a draft agreement, we reserve the right to modify it.

3) We reserve the right to revoke this delegation at any time.

Part 5 Extensive Reading

(A)

1.

1) 钢筋砼墙

2) 砼基脚

3) 钢筋

4) 浇筑混凝土

5) 给模板内浇筑混凝土

6) 将混凝土夯实

7) 给模板上部铺上塑料布

8) 让混凝土处于养护状态

9) 从墙上拆下模板

10) 修补墙上的大缺陷

2.

1) F 2) F 3) F 4) T 5) T 6) F 7) T 8) F

(B)

1.

1) B 2) A 3) D 4) C 5) A 6) D 7) A 8) A

2.

1) In the evolution of structural systems 2 basic types can be distinguished, deriving at various times according to particular circumstances. The systems are massive structures and skeleton structures.

247

2) All non-load bearing walls, adopted at any time by whatever structural tradition, could in a certain sense be called curtain walls, In fact, in some respects, they possess both the qualities of a wall(which is immovable, heavy and definitive) and those of a curtain(which is movable, light and temporary).

3) This definition, however, is not always strictly adhered to, either, because curtain walling is of recent invention, and has thus not been the subject of any deep critical examination, or because its typology is being constantly enriched, thanks to the continual efforts of designers and manufacturers to find better methods of production and application.

4) Between 1850 and the early years of the 20th century, the window gradually turned into the window wall, sometimes taking over the whole basic area defined by the facade. The use of large areas of glass had meanwhile become widespread in greenhouses and winter gardens, pedestrian galleries, railway station roofs and large exhibition pavilions.

5) The small dimensions of the steel frames and the progress made by the glass industry permitted an increase in window sizes, a development that was also stimulated by the demand for as much natural light as possible in industrial and commercial buildings. The transformation of the window into the window wall and the employment of large glazed areas drew attention to a number of problems and evoked the first solutions to them. This included questions of insulation, eliminating condensation, developing the secondary glazing framework and considering the effects of expansion by carefully designing joints and casings.

6) Between the eve of the First World War and the beginning of the Second, the architects of the modern movement carried out a series of experiments, each of which may be considered as perfecting some particular aspect of curtain walling by the use of modern methods of industrial production. At the same time, the theoretical principles were formulated by new architectural schools such as the "Bauhaus" in Germany and various projects by European architects such as Mies van der Roche and Walter Gropius. It was only after the Second World War, however, that the first built experiments emerged on a vast scale. It was in this period that the curtain wall became that popular in the United States and Europe.

Unit 7

Part 1 Warm-up Activities

1. 水平屋顶；斜坡屋顶；人字形屋顶；四坡屋顶；复折式屋顶；复斜屋顶

2.

1) Although termed flat roofs, they are often constructed with a slight fall to enable the rainwater to run off.

2) The main advantages of flat roofs are that they are comparatively simple to construct and generally less costly than pitched roofs.

Keys to the Exercises from Unit 1 to Unit 14

3.

1) No. The steeper the pitch, the more effective the roof is in quickly disposing of rainwater or snow. On the other hand, a steeper pitch entails a larger roof area and a higher cost.

2) The steeper the pitch, the more effective the roof is in quickly disposing of rainwater or snow. On the other hand, a steeper pitch entails a larger roof area and a higher cost.

Part 2 Controlled Practices

1.

a) to fix the timber bearers

b) to cover the timber bearers with timber boarding

c) to nail the first layer of tar into the boarding

d) the subsequent layers are bonded to each other in hot tar

e) The felt is then covered with stone chippings

2.

①墙面砖；

②外墙面；

③通风口；

④镶边板；

⑤压顶石；

⑥金属防雨板；

⑦2～3层沥青层；

⑧粗面企口咬合板；

⑨屋顶承木；

⑩接合料；

⑪碎屑层；

⑫隔温(水)层；

⑬木板条；

⑭舌槽企口板；

⑮内墙面；

⑯抹灰层。

3. ①山墙；②山墙檐口；③山墙顶点；④立柱；⑤烟囱管道；⑥对角撑；⑦大梁(脊檩)；⑧烟囱；⑨烟道；⑩檩条；⑪椽子；⑫承梁板；⑬托梁；⑭屋檐

4.

a) fix ridge purlins.

a) fix strut to each side of the posts to provide further support to the ridge purlins.

b) nail the rafters, spaced at approximately 80ctn centers, to the purlins.

d) fix eaves purlins over the ceiling joists.

Part 3 Further Development

1. ①承梁板；②立柱；③对角撑；④螺栓啮接；⑤（楔栓）加固；⑥椽子；⑦屋脊板；⑧屋脊结点联结板；⑨连接梁；⑩檩条；⑪对角撑；⑫短粗榫舌；⑬托梁；⑭垫片

Part 4 Business Activities

1）我对每台木工锯刨都装设有保护罩感到满意，这利于工人的安全。不过没有足够的消防设施。你们在这车间必须增加至少10台灭火器。而且应在墙壁上挂上许多诸如"禁止吸烟"的标牌。

2）这是我们的机械维修间。它分为两部分：一部分用于修理重型机械，如挖掘机、推土机、履带吊、混凝土泵等等；另一部分用于修理运输工具如卡车和客车。龙门吊起重重量为5吨。该汽车修理间装备比街上的修理铺好得多。

3）但这仓库用来储存水泥并不好。首先，由于楼层高度不足以保持干燥，楼面是湿的；其次，屋顶防水不够。你们可以看见屋顶有一个孔，阳光可以穿透进来。

Part 5 Extensive Reading

(A)

1) C 2) A 3) A 4) C 5) C 6) B 7) C 8) C

(B)

1.

1) 建筑物的覆盖部分
2) 防风、防雨、防雪
3) 在住宅里
4) 与屋面形成相应的角度
5) 在酸性环境中的耐久性
6) 现代办公大楼
7) 木构架
8) 质量合格的雨水槽浇筑在混凝土中
9) 横向构件
10) 印染厂和漂白厂
11) 使设计具有美学价值
12) 由内柱支撑
13) 在单坡结构中
14) 在每根椽子的中部支一根立柱
15) 水平系梁
16) assemble and erect
17) protect the upper portion of the external walls
18) slope roofing materials
19) at the eaves
20) a membrane
21) a large gabled roof structure
22) light supporting beams
23) special felt asbestos sheets
24) timber trusses with strong connections
25) modern roofing materials

2.

1) A roof's chief purpose is to enclose the upper parts of a building as a protection against wind, rain and snow.

2) The pitch roof has to give way to the flat roof for the construction of the flat roof

is cheaper than the pitch roof.

3) Steel framework can be prepared, assembled and erected with rapidity.

4) The method of support for very large roofs is the vertical support provided at an intermediate point of either rafter.

5) Steel or reinforced or prestressed concrete is commonly used in gabled roofs with spans of 200 feet(60 meters) or more.

Unit 8

Part 1 Warm-up Activities

1. lump hammer 大锤　bolster 托板　trowel 泥刀　shovel 锹　float 镘刀　plane 刨子　panel saw 手锯　spanner 扳子　brace and bit 手工钻和钻头　mallet 木槌　chisel 凿子　pincers 夹钳　vice and file 老虎钳和锉刀　screwdriver 螺丝刀　cable shear 钢筋剪　wire stripper 剥线钳 combination pliers 多用钳　brush 刷子

2.

a) A brace and bit is a tool for drilling holes in wood.

b) A shovel is a tool for mixing mortar.

c) A float is a tool for smoothing the plaster on a wall.

d) A panel saw is a tool for cutting wood.

e) A vice and hacksaw is a tool for cutting metal pipes.

f) A cable shears is a tool for cutting electric cables.

g) A mallet and chisel is a tool for making joint.

h) A file is a tool for smoothing metal surfaces.

i) A wire strippers is a tool for removing the outer sheathing of wire.

j) A screw driver is a tool for turning screws.

k) A brush is a tool for painting surfaces.

l) A lump hammer and bolster is a tool for cutting bricks.

m) A spanner is a tool for tightening nuts.

n) A combination pliers is a tool for twisting strands of wire together.

o) A plane is a tool for smoothing wood surfaces.

p) A trowel is a tool for laying mortar on bricks.

q) A pincer is a tool for removing nails.

4.

A lighting engineer uses a daylight factor meter to measure the illumination from the sky.

A structural engineer uses a hygrometer to measure the relative humidity.

A structural engineer uses a thermometer to measure the temperature.

A structural engineer uses a strain gauge to measure the strain on a structure.

A bricklayer uses a plumb-bob to check the verticality.

A bricklayer uses a steel tape to measure distances.

A bricklayer uses a spirit level to check vertical and horizontal work.

An acoustic engineer uses a sound pressure meter to measure the sound pressure.

A carpenter uses a square to check squareness.

An electrician uses a voltmeter to measure the voltage of a circuit.

Part 2　Controlled Practices

2. We can conclude that the strength of concrete is considerably reduced as a result of the additional water. The reason for this is that water combines chemically with cement and an excess of water weakens this reaction on which the strength of the concrete depends.

Part 4　Business Activities

1. "quote sb. (the price/rate) for sth. ""请给……报……的价格/单价。"

1)A：Our company deals in various construction materials.

B：Please quote us the rate for all-in aggregate.

2)A：Please quote us for this kind of mosaic.

3)B：Please quote us the lump sum for one hundredset.

2. "be in urgent need of sth. "急需……"

1)A：We're in urgent need of 10 tons of galvanized steel sheet. I wonder if you could deliver it within one week.

2)B：It is true that we are in urgent need of it, but I'm afraid the price is too high for us to accept.

3)A：This disease is contagious and the patient is in urgent need of medical treatment.

3. "reduce the price/rate by..." "把价格/费率降低……"

1)B：All right. We'll reduce the price by 3 percent for your order this time.

2)A：If you don't reduce this price by 10 dollars, we have to make the purchase from other manufacturers.

3)A：As a rule, we reduce the price by 5 percent to regular customers. I hope you could come often.

Part 5　Extensive Reading

(A)

a) To know the ratio of the light in the room to the light of the unobstructed sky.

b) from the sky; reflected light of external surfaces; light received by reflection from the internal surfaces of the room

c) The direct light from the sky which reaches any given point in a room is determined by how big a patch of sky can be seen from that point, or, more strictly, the projected solid angle subtended by the patch of visible sky at that point. It is also determined by the brightness of the patch of sky.

d) If the brightness of the patch of sky can be assumed to be uniform, the ratio of direct internal light to the external light from the sky is known as the sky component, and it is proportional to this projected solid angle.

Keys to the Exercises from Unit 1 to Unit 14

e) A simpler method of determining the direct light from the sky is by means of sky component protractors.

f) It can be laid directly on to the working drawings.

(B)

1. 1) C 2) A 3) A 4) D 5) B 6) B 7) C 8) A
2. 1) F 2) F 3) T 4) F 5) F 6) F 7) F 8) F

Unit 9

Part 1 Warm-up Activities

3.

a) concrete floor slabs

b) horizontal cladding panels

c) phase 2

d) column base plates

e) phase 4

f) corner units

Part 2 Controlled Practices

3.

a) Trade: <u>plumbers</u>

Job: <u>installing the pipe work and fitting stationary</u>

Weeks working: <u>21 to 30</u>

Trade: <u>carpenters</u>

Job: <u>fixing floor joists, roof timbers, doors, windows, etc.</u>

Weeks working: <u>30 to 44</u>

Trade: <u>roofing contractor</u>

Job: <u>laying roof covering</u>

Weeks working: <u>21 to 35</u>

b) Trade: <u>laborers</u>

Job: <u>doing manual work</u>

Weeks working: <u>0 to 55</u>

Job: <u>machine drivers</u>

Trade: <u>excavating ground</u>

Weeks working: <u>0 to 9</u>

Trade: steel erectors
Job: erecting the steelwork
Weeks working: 9 to 15

c) Trade: heating contractor
Job: installing heating equipment
Weeks working: 25 to 40

Trade: decorators
Job: decorating building
Weeks working: 44 to 50

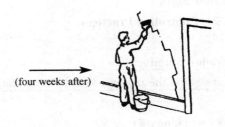

(four weeks after)

Trade: electricians
Job: installing electrical equipment
Weeks working: 25 to 40

4.
a) brick layers
b) heating contractors and electricians
c) plumbers and cladding fixers
d) decorators
e) machine drivers

Part 4　Business Activities

Now try to do the following translation:

1. "apply(to...)for..." is used frequently in construction management, "(向……)申请……"

1) B: In most cases, yes. However, you have to apply to the Labor Department for their work permits and give the reasons.

2) B: Thank you. I hope you can assist us in applying for the registration of our trade mark in your country.

3) A: Please send us all the documents by Express mail so that we can apply for the import license.

2. "What if...?"

1) B: What if there's no direct flight?

2) A: What if it rains tomorrow?

3) A: What if the steel bars get rusty?

3. "draw one's attention to sth."

1) B: I'd like to draw your attention to the variation order issued by you on May 2.

2) B: May I draw your attention to our response dated June 8?

3) A: Last but not the least, I'd like to draw your attention to this milestone date.

Part 5　Extensive Reading

(A)

1) 工地劳动力　2) 建筑工地（施工现场）　3) 拥有承包商权限　4) 工程委托方（甲方）耗资高　5) 放线　6) 使承包合同按时完成　7) 某一工种的工长　8) 给建筑物的主要位置放线　9) 制作混凝土模板　10) 采用特定的顺序

11) a contractor

12) the contractor's agent

13) the resident engineer

14) set out for the other trades

15) a clerk of the works; a civil engineering inspector

16) see to the arrival of the essential material

17) set out the formwork for carpenter

18) be sure that nothing is forgotten; be sure to forget nothing

19) the progress of the job

20) follow the design as it is laid down in the drawings

(B)

1) F　2) T　3) T　4) F　5) F　6) T　7) T

Unit 10

Part 1　Warm-up Activities

2.

a) doing experiment

b) library

c) having dinners

d) sleeping

e) kitchen

f) providing a therapy that cures disease

g) dispensing medicines

h) examination room

i) storehouse

j) mixing concrete

Part 2 Controlled Practices

3.

	Element	Main functions
External envelope	lowest floor	C, H, J, etc.
	external wall	I H F A B D E G
	roof	I H A E G
Internal division	suspended floor	I H B C E
	partitions	I H F B E
	suspended ceiling	K E

4. a)—lowest floor

　　—external wall and partition

　　—roof and external wall

　　—partition

　　—artificial light

　　—external wall

b)—partition

　　—roof

　　—lowest floor

　　—external wall

　　—external wall

c)—lowest floor, external wall and roof

　　—roof, external wall and partitions

　　—partition and suspended floor

　　—roof and external wall

　　—lowest floor and external wall

6. 1) steel　2) concrete　3) mineral wool　4) concrete　5) concrete　6) ceramic tile

Part 4 Business Activities

1. 首先，根据施工监督体系对工地安全进行定期检查，确保作业过程采用安全的施工方法和遵守安全健康条例。

2. 安全负责人的第二个职责就是对于事故和危险事件进行严密的调查，确定起因，制定有效的解决方法，防止类似事件再次发生。

3. 培训内容涵盖多个方面，例如工地安全措施、中国劳动法规定的雇员责任、公司安全制度、建筑事故和报告、用电安全、个人安全设施、起重机安全、呼吸防护、防火和消防、有毒物品、急救和紧急援助程序，等等。

4. 由于它用于外墙施工，我们对于脚手架进行了全面的检查，包括基底与墙面的连接点、走道、扶手、支撑杆、阶梯,并进行了承重试验,等等。

Part 5　Extensive Reading

(A)

1. 1) D　2) A　3) B　4) B　5) C　6) C　7) A　8) C　9) A　10) D

2. 1) The minimum possible distance traveled by the users of the building.

2) It will result in additional cost to the construction and other items such as monthly utility bills, maintenance, etc.

3) The latter.

4) Orient the short dimension of the room perpendicular to the hall.

5) Remove the wall between the two.

6) Walls and locked doors are used to achieve the goal or a receptionist is set up as a barrier.

7) Group the bedrooms together and separate them from the public areas.

8) Bring them together to reduce the cost of piping and the labor to install it and stack rooms having plumbing in multistory buildings.

(B)

1) (A—B)　2) (D)　3) (C)　4) (F)　5) (E)

Unit 11

Part 2　Controlled Practices

3. a) kg　b) kg/m³　c) dB　d) N/mm²　e) lx　f) lm　g) A　h) ℃　i) J

Part 3　Further Development

1. a) Celsius　b) stress　c) decibel　d) lux　e) mass　f) ampere　g) illumination　h) Watts

3.

a) There are <u>five</u> people. The floor area allowed per person is <u>10 square meters</u>. Therefore <u>the floor area required is at least 50 square meters</u>.

b) The load on the column is <u>2,000 Newton</u>. The <u>compressive stress</u> allowed in the concrete is <u>5 N/mm²</u>. Therefore <u>the minimum cross-sectional area of the column required is 400mm²</u>

c) The volume of <u>the concrete wall is 10 cubic meters. The maximum weight of the wall allowed is 22,000 kilogrammes</u>. Therefore <u>the maximum density of concrete required is 2,200kg/m³</u>.

Part 4　Business Activities

1.

1) B: Could you elaborate on your specific measures?

2) B: We hope you can elaborate on it.

3) A: You just tell us the facts. Don't elaborate on them.

2.

1) You are welcome to visit our company.

2) You are welcome to give us your suggestions.

3) You are welcome to use this equipment.

Part 5　Extensive Reading

(A)

1.

1) F　Concrete is made from four different materials: cement, coarse aggregate (stones), fine aggregate(sand or crushed stone) and water.

2) F　Coarse aggregate range in size from 5mm to 40mm for normal work.

3) T

4) T

5) T

6) T

7) F　When the minimum horizontal distance between reinforcing rods is 15mm, the maximum size of aggregate should be 10mm.

2.

1) Yes, Concrete is a very strong material when it is placed in compression. But it is extremely weak in tension.

2) We use reinforcement in concrete structures.

3) There are many ways to test the strength of a batch of concrete. The tests used can be categorized as destructive and nondestructive tests.

4) These cylinders are cured for 28 days.

5) These cylinders are tested by compression until they are crushed.

6) This will give the contractor or the engineer the compressive strength for that batch of concrete.

(B)

1.

1) A　2) B　3) D　4) A　5) D　6) C　7) B　8) B

2.

1) as well as　2) take into account　3) by means of　4) as though　5) perpendicular to　6) are equipped with　7) based on　8) parallel to

Unit 12

Part 2　Controlled Practices

1. 1) extending　2) excessive　3) comfort　4) shade　5) screening　6) occur　7) radiation　8) adequate　9) interior　10) consequently

2. Type A is an open layout, and it is suitable for warm-humid climates. According to the passage given above, in warm-humid climate, the rooms of houses must have

258

adequate shade and ventilation. Usually houses have an open layout to gain maximum benefit from the prevailing wind and walls have less importance.

Type B is a compact layout and it is suitable for hot-dry climates, because in this climate houses must give adequate protection against the excessive heat of the sun. Usually they have compact layouts, so that surfaces exposed to the sun are reduced as much as possible. Thick Walls with heat storing materials can hold the heat of the day and give it back to the interior of the house at night.

3. In composite climates, houses designed to perform <u>adequately</u> for one season will perform <u>inadequately</u> for the other. To solve this problem, houses are sometimes built two storeys high. The ground floor is built with <u>excessively</u> thick walls. These retain the heat so that it is <u>warm enough</u> to sleep comfortably on the ground floor during the <u>coolest</u> of the year. The first floor structure is built with materials. This structure cools quickly at night so that it is to sleep comfortably on the first floor during the <u>hottest</u> part of the year.

Part 4 Business Activities

1.
1) B: May I refer you to Clause 20 of the contract?
2) B: May I refer you to the minutes of the meeting held on August 9th?
3) B: May I refer you to the stop-work order issued by your resident engineer on July 8th?

2.
1) B: There's something in what you say; but what about the extra cost?
2) B: There's something in what you say; but will it affect the functions of the whole facility?
3) B: There's something in what you say; but do they have the required skills?

3.
1) You're instructed to suspend the shipment.
2) You're instructed to stop erecting the formwork.
3) You're instructed to resume the work.

Part 5 Extensive Reading

(A)

1. 1) 地席的尺寸 2) 面朝阳的窗户 3) 日本房屋的突出特征 4) 与当地的风景相吻合 5) 造几间有充足阳关的卧室 6) 中国古建筑的特点 7) 社会习俗 8) 房屋的外貌与地方条件相适应

2. 1) balcony; courtyard 2) flat; including 3) shutter; glare 4) shaded; make out 5) local; landscape; texture 6) dwellings 7) veranda

(B)

1.
1) F 2) T 3) F 4) F 5) T 6) F 7) T

259

2.

1) exposes   2) extreme   3) immune   4) purpose   5) retained   6) attributable   7) purpose   8) contributed   9) resistance   10) component

Unit 13

Part 1 Warm-up Activities

2. a) water tower   b) tropical house   c) columns   d) columns   e) 1 to 3   f) 3 to 5   g) tropical house   h) 2 to 1   i) 6 to 1   j) block of flats

3. For example:

a) Compared with a water tower, a micro-wave tower supports relatively light load and has a proportionately thinner tower structure.

b) In comparison with a tropical house, a block of flats support relatively heavy load and has a proportionately shorter columns.

4.

a) are relatively long and thin in proportion to its size; lighter one needs proportionately longer and thinner columns

b) thickness; length; thicker; shorter; the thinner and longer its columns

Part 2 Controlled Practices

1. a) T

b) F   The structure of the water tower has to support more weight than that of the micro-wave tower.

c) T

d) F   The strength of a column is directly proportional to its thickness and inversely proportional to its height.

e) F   Compare with a water tower, a micro-wave tower has a relatively tall structure.

f) F   The lighter the load on a tower, the thinner its structure.

g) F   Similarly, the heavier a building, the thicker its columns.

2. a) $9m^2$ b) $12m$ c) $4/3$ d) $36m^2$ e) $24m$ f) $2/3$ g) smaller   h) longer

3. a) $56m$; $70m$ b) rectangular building; compact shape; longest   c) compare; find; the smaller perimeter in proportion to area   d) conclude; perimeter/area; the size; the shape

Part 3 Further Development

2. a)

e. g. 1) Heat transfer is directly proportional to surface area

2) Heat transfer is inversely proportional to volume.

3) Heat loss is directly proportional to air temperature gradient.

4) Heat loss is inversely proportional to thickness of insulation.

b)

e. g. 1) The higher the ratio between surface area and volume, the more quickly it regains or loses heat.

2) The higher the ratio between surface area and volume, the fast the rate of heat transfer.

3) The thicker the insulation of a building, the slower the rate of heat transfer.

Part 4　Business Activities

1.

1)B: I'd prefer to stay indoors to prepare the document for tomorrow's negotiation if you don't mind.

2)B: I want to leave it to be discussed in our next meeting if you don't mind.

3)B: I'd like to have a stroll in the Palace Museum if you don't mind.

2.

1)A: Are you finished with your work?

2)A: When do you think you will be finished with the river closure?

Part 5　Extensive Reading

(A)

1. 1) B　2) A　3) D　4) C

2. 1) A　2) D　3) C　4) B

(B)

1.

1)未来有很多的惊喜在等待着我们。

2)虽然在专家们眼中还只是一线亮光,但是我们已经在考虑它的可能性了。

3)利用配有特殊的防辐射设备和方法,核技术将用于大型项目的挖掘。

4)预制件将更加先进,基于公制模块单位的统一标准化规范将被广泛应用。

5)许多设备的日常功能将成为自动化或由计算机控制,尤其是在住宅建筑。

6)照明、采暖、厨房设施以及许多其他设备将会定时运作以满足个体要求,其完整的综合程序能够通过电话远程控制被触发或调整。

2. 1)T　2) F　3)F　4)T　5)F　6)T　7)T　8)T

Unit 14

Part 1　Warm-up Activities

1.

e. g. 1) The column moved with the result, that/and as a result, the span increased.

2) The span increased with the result that/and as a result the beam deflected excessively.

3) The beam deflected excessively with the result that/and as a result the cracks formed on the underside of beam

3. a) The presence of soluble salts in the hardcore resulted in/caused its the interaction with cement in concrete floor. This, in turn, caused the expansion of concrete floor and the formation of cracks in the floor.

b) Gravel being washed away by heavy rain resulted/caused that the roof is inadequately protected from the sun. This, in turn, caused the roof heated up and the failure of the wall/roof joint.

c) The poor quality paint resulted in/caused the increase of moisture content of wood. This, in turn, caused the drying and contraction of the wood and formation of gaps between the window and the frame.

Part 2　Controlled Practices

1.

a) prevents the floor from cracking

b) prevents the wall and roof joint from failure

c) prevents the gaps between window and frame woodwork from expanding

3.

a) Yes, because the loads from the internal partitions of a building rarely change with time and are fixed in one place, and they are dead loads.

b) No, because loads from storage in a building are always variable and movable, and they are live loads.

c) Engineers can only predict from past experience.

d) Because we can never know the exact value of either the applied loads or the strength of the structure.

e) A high safety factor.

f) When the safety factor is below 1.

Part 3　Further Development

2.

a) The wind is directed into the basement because on the flat roof there is a tower facing the prevailing wind.

b) The internal party wall is kept cool because the wall is very thick and does not receive any direct solar radiation.

c) The incoming air comes into contact with the surfaces of the duct and, as a result, is cooled by conduction.

d) The relative humidity of the air is increased because it passes over porous water jugs just before it enters the basement.

e) The courtyard is ventilated during the daytime when the cooling air leaves the basement through an outlet thereby.

3.

a) T

b) T

c) F　Frequent painting prevents steelwork from corrosion.

d) F　Insulating a house allows the air inside to be cool or warm to a certain temperature or humidity.

e) F　The expansion of a roof is caused by an increase in temperature.

f) F   In an arch the bricks are wedge-shaped thereby causing their weight to be distributed downwards along the curve of the arch.

g) T

h) T

i) F   The contraction of a column is due to pressure forces.

j) T

k) T

Part 4 Business Activities

1. louvers   2. ventilator   3. retention money   4. "as built" drawings   5. specification   6. progress payment   7. erection   8. procurement

Part 5 Extensive Reading

(A)

1) The fatigue crack forms at the surface and progresses into the body of the material.

2) B   3) D   4) A   5) D

(B)

1. 1) A   2) D   3) A   4) A   5) A   6) B   7) C   8) C

2. 1) F   2) T   3) T   4) F   5) F   6) F   7) T   8) T

Appendix
Vocabulary and Expressions of Civil Engineering & Architecture
(常用土木与建筑工程专业词汇)

A

acceptable quality 合格质量

acceptance lot 验收批量

acierate 钢材

admixture 外加剂

against slip coefficient between friction surface of high-strength bolted connection 高强度螺栓摩擦面抗滑移系数

aggregate 骨料

air content 含气量

air-dried timber 气干材

allowable ratio of height to sectional thickness of masonry wall or column 砌体墙、柱容许高比

allowable slenderness ratio of steel member 钢构件容许长细比

allowable slenderness ratio of timber compression member 受压木构件容许长细比

allowable stress range of fatigue 疲劳容许应力幅

allowable ultimate tensile strain of reinforcement 钢筋拉应变限值

allowable *value* of crack width 裂缝宽度容许值

allowable *value* of deflection of structural member 构件挠度容许值

allowable *value* of deflection of timber bending member 受弯木构件挠度容许值

allowable *value* of deformation of steel member 钢构件变形容许值

allowable *value* of deformation of structural member 构件变形容许值

allowable *value* of drift angle of earthquake resistant structure 抗震结构层间位移角限值

amplified coefficient of eccentricity 偏心距增大系数

anchorage 锚具

anchorage length of steel bar 钢筋锚固长度

approval analysis during construction stage 施工阶段验算

arch 拱

arch with tie rod 拉捍拱

arch-shaped roof truss 拱形屋架

area of shear plane 剪面面积

area of transformed section 换算截面面积

aseismatic design 建筑抗震设计

assembled monolithic concrete structure 装配整体式混凝土结构

automatic welding 自动焊接

auxiliary steel bar 架立钢筋

B

backfilling plate 垫板

balanced depth of compression zone 界限受压区高度

balanced eccentricity 界限偏心距

bar splice 钢筋接头

bark pocket 夹皮

batten plate 缀板

beam 次梁

bearing plane of notch 齿承压面

bearing plate 支承板

bearing stiffener 支承加劲肋

bent-up steel bar 弯起钢筋

block 砌块

block masonry 砌块砌体

block masonry structure 砌块砌体结构

blow hole 气孔

board 板材

bolt 螺栓

bolted connection(钢结构)螺栓连接

bolted joint(木结构)螺栓连接

bolted steel structure 螺栓连接钢结构

bonded prestressed concrete structure 有黏结预应力混凝土结构

bow 顺弯

brake member 制动构件

breadth of wall between windows 窗间墙宽度

brick masonry 砖砌体

brick masonry column 砖砌体柱

brick masonry structure 砖砌体结构

brick masonry wall 砖砌体墙

broad-leaved wood 阔叶树材

building structural materials 建筑结构材料

building structural unit 建筑结构单元

building structure 建筑结构

built-up steel column 格构式钢柱
bundled tube structure 成束筒结构
burn-through 烧穿
butt connection 对接
butt joint 对接
butt weld 对接焊缝

C

calculating area of compression member 受压构件计算面积
calculating overturning point 计算倾覆点
calculation of load-carrying capacity of member 构件承载能力计算
camber of structural member 结构构件起拱
cantilever beam 挑梁
cap of reinforced concrete column 钢筋混凝土柱帽
carbonation of concrete 混凝土碳化
cast-in-situ concrete slab column structure 现浇板柱结构
cast-in-situ concrete structure 现浇混凝土结构
cavitation 孔洞
cavity wall 空斗墙
cement 水泥
cement content 水泥含量
cement mortar 水泥砂浆
characteristic *value* of live load on floor or roof 楼面、屋面活荷载标准值
characteristic *value* of wind load 风荷载标准值
characteristic *value* of concrete compressive strength 混凝土轴心抗压强度标准值
characteristic *value* of concrete tensile strength 混凝土轴心抗拉强度标准值
characteristic *value* of cubic concrete compressive strength 混凝土立方体抗压强度标准值
characteristic *value* of earthquake action 地震作用标准值
characteristic *value* of horizontal crane load 吊车水平荷载标准值
characteristic *value* of masonry strength 砌体强度标准值
characteristic *value* of permanent action 永久作用标准值
characteristic *value* of snow load 雪荷载标准值
characteristic *value* of strength of steel 钢材强度标准值
characteristic *value* of strength of steel bar 钢筋强度标准值
characteristic *value* of uniformly distributed live load 均布活荷载标准值
characteristic *value* of variable action 可变作用标准值
characteristic *value* of vertical crane load 吊车竖向荷载标准值
characteristic *value* of material strength 材料强度标准值
checking section of log structural member 原木构件计算截面

chimney 烟囱

circular double-layer suspended cable 圆形双层悬索

circular single-layer suspended cable 圆形单层悬索

circumferential weld 环形焊缝

classification for earthquake-resistance of buildings 建筑结构抗震设防类别

clear height 净高

clincher 扒钉

coefficient of equivalent bending moment of eccentrically loaded steel member(beam-column) 钢压弯构件等效弯矩系数

cold bend inspection of steel bar 冷弯试验

cold drawn bar 冷拉钢筋

cold drawn wire 冷拉钢丝

cold-formed thin-walled section steel 冷弯薄壁型钢

cold-formed thin-walled steel structure 冷弯薄壁型钢结构

cold-rolled deformed bar 冷轧带肋钢筋

column bracing 柱间支撑

combination *value* of live load on floor or roof 楼面、屋面活荷载组合值

compaction 密实度

compliance control 合格控制

composite brick masonry member 组合砖砌体构件

composite floor system 组合楼盖

composite floor with profiled steel sheet 压型钢板楼板

composite mortar 混合砂浆

composite roof truss 组合屋架

composite member 组合构件

compound stirrup 复合箍筋

compression member with large eccentricity 大偏心受压构件

compression member with small eccentricity 小偏心受压构件

compressive strength at an angle with slope of grain 斜纹承压强度

compressive strength perpendicular to grain 横纹承压强度

concentration of plastic deformation 塑性变形集中

conceptual earthquake-resistant design 建筑抗震概念设计

concrete 混凝土

concrete column 混凝土柱

concrete consistence 混凝土稠度

concrete folded-plate structure 混凝土折板结构

concrete foundation 混凝土基础

concrete mix ratio 混凝土配合比

concrete wall 混凝土墙

concrete-filled steel tubular member 钢管混凝土构件

conifer 针叶树

coniferous wood 针叶树材

connecting plate 连接板

connection 连接

connections of steel structure 钢结构连接

connections of timber structure 木结构连接

consistency of mortar 砂浆稠度

constant cross-section column 等截面柱

construction and examination concentrated load 施工和检修集中荷载

continuous weld 连续焊缝

core area of section 截面核芯面积

core tube supported structure 核心筒悬挂结构

corrosion of steel bar 钢筋锈蚀

coupled wall 耦合壁

coupler 连接器

coupling wall-beam 连梁

coupling wall-column... 墙肢

coursing degree of mortar 砂浆分层度

cover plate 盖板

covered electrode 焊条

crack 裂缝

crack resistance 抗裂度

crack width 裂缝宽度

crane girder 吊车梁

crane load 吊车荷载

creep of concrete 混凝土徐变

crook 横弯

cross beam 井字梁

cup 翘弯

curved support 弧形支座

cylindrical brick arch 砖筒拱

D

decay 腐朽

decay prevention of timber structure 木结构防腐

defect in timber 木材缺陷

deformation analysis 变形验算

degree of gravity vertical for structure or structural member 结构构件垂直度

degree of gravity vertical for wall surface 墙面垂直度

degree of plainness for structural member 构件平整度
degree of plainness for wall surface 墙面平整度
depth of compression zone 受压区高度
depth of neutral axis 中和轴高度
depth of notch 齿深
design of building structures 建筑结构设计
design *value* of earthquake-resistant strength of materials 材料抗震强度设计值
design *value* of load-carrying capacity of members 构件承载能力设计值
designations of steel 钢材牌号
design *value* of material strength 材料强度设计值
destructive test 破损试验
detailing reinforcement 构造配筋
detailing requirements 构造要求
diamonding 菱形变形
diaphragm 横隔板
dimensional errors 尺寸偏差
distribution factor of snow pressure 屋面积雪分布系数
dog spike 扒钉
double component concrete column 双肢柱
dowelled joint 销连接
down-stayed composite beam 下撑式组合梁
ductile frame 延性框架
dynamic design 动态设计

E

earthquake-resistant design 抗震设计
earthquake-resistant detailing requirements 抗震构造要求
effective area of fillet weld 角焊缝有效面积
effective depth of section 截面有效高度
effective diameter of bolt or high-strength bolt 螺栓(或高强度螺栓)有效直径
effective height 计算高度
effective length 计算长度
effective length of fillet weld 角焊缝有效计算长度
effective length of nail 钉有效长度
effective span 计算跨度
effective supporting length at the end of beam 梁端有效支承长度
effective thickness of fillet weld 角焊缝有效厚度
elastic analysis scheme 弹性方案
elastic foundation beam 弹性地基梁
elastic foundation plate 弹性地基板

elastically supported continuous girder 弹性支座连续梁
elasticity modulus of materials 材料弹性模量
elongation rate 伸长率
embedded parts 预埋件
enhanced coefficient of local bearing strength of materials 局部抗压强度提高系数
entrapped air 含气量
equilibrium moisture content 平衡含水率
equivalent slenderness ratio 换算长细比
equivalent uniformly distributed live load 等效均布活荷载
effective cross-section area of high-strength bolt 高强度螺栓的有效截面积
effective cross-section area of bolt 螺栓有效截面面积
euler's critical load 欧拉临界力
euler's critical stress 欧拉临界应力
excessive penetration 塌陷

F

fiber concrete 纤维混凝土
filler plate 填板门
fillet weld 角焊缝
final setting time 终凝时间
finger joint 指接
fired common brick 烧结普通砖
fish eye 白点
fish-belly beam 角腹式梁
fissure 裂缝
flexible connection 柔性连接
flexural rigidity of section 截面弯曲刚度
flexural stiffness of member 构件抗弯刚度
floor plate 楼板
floor system 楼盖
four sides(edges)supported plate 四边支承板
frame structure 框架结构
frame tube structure 单框筒结构
frame tube structure 框架-筒体结构
frame with sides way 有侧移框架
frame without sides way 无侧移框架
flange plate 翼缘板
friction coefficient of masonry 砌体摩擦系数
full degree of mortar at bed joint 砂浆饱满度
function of acceptance 验收函数

G

gang nail plate joint 钉板连接
glue used for structural timber 木结构用胶
glued joint 胶合接头
glued laminated timber 层板胶合木
glued laminated timber structure 层板胶合结构
grider 主梁
grip 夹具
grith weld 环形焊缝
groove 坡口
gusset plate 节点板

H

hanger 吊环
hanging steel bar 吊筋
heartwood 心材
heat tempering bar 热处理钢筋
height variation factor of wind pressure 风压高度变化系数
high-strength bolt 高强度螺栓
high-strength bolt with large hexagon bolt 大六角头高强度螺栓
high-strength bolted bearing type join 承压型高强度螺栓连接
high-strength bolted connection 高强度螺栓连接
high-strength bolted friction-type joint 摩擦型高强度螺栓连接
high-strength bolted steel structure 高强螺栓连接钢结构
hinge support 铰轴支座
hinged connection 铰接
hingeless arch 无铰拱
hollow brick 空心砖
hollow ratio of masonry unit 块体空心率
honeycomb 蜂窝
hook 弯钩
hoop 箍筋
hot-rolled deformed bar 热轧带肋钢筋
hot-rolled plain bar 热轧光圆钢筋
hot-rolled section steel 热轧型钢
hunched beam 加腋梁

I

impact toughness 冲击韧性
impermeability 抗渗性
inclined section 斜截面

inclined stirrup 斜向箍筋
incomplete penetration 未焊透
incomplete fusion 未熔合
incompletely filled groove 未焊满
indented wire 刻痕钢丝
influence coefficient for load-bearing capacity of compression member 受压构件承载能力影响系数
influence coefficient for spacial action 空间性能影响系数
initial control 初步控制
insect prevention of timber structure 木结构防虫
inspection for properties of glue used in structural member 结构用胶性能检验
inspection for properties of masonry units 块体性能检验
inspection for properties of mortar 砂浆性能检验
inspection for properties of steel bar 钢筋性能检验
integral prefabricated pre-stressed concrete slab-column structure 整体预应力板柱结构
intermediate stiffener 中间加劲肋
intermittent weld 断续焊缝

J

joint of reinforcement 钢筋接头

K

key joint 键连接
kinetic design 动态设计
knot 节子（木节）

L

laced of battened compression member 格构式钢柱
lacing and batten elements 缀材（缀件）
lacing bar 缀条
lamellar tearing 层状撕裂
lap connection 叠接（搭接）
lapped length of steel bar 钢筋搭接长度
large panel concrete structure 混凝土大板结构
large-form concrete structure 大模板结构
lateral bending 侧向弯曲
lateral displacement stiffness of storey 楼层侧移刚度
lateral displacement stiffness of structure 结构侧移刚度
lateral force resistant wall-structure 抗侧力墙体结构
leg size of fillet weld 角焊缝焊脚尺寸
length of shear plane 剪面长度

lift-slab structure 升板结构

light weight aggregate concrete 轻骨料混凝土

limit of acceptance 验收界限

limiting *value* for local dimension of masonry structure 砌体结构局部尺寸限值

limiting *value* for sectional dimension 截面尺寸限值

limiting *value* for supporting length 支承长度限值

limiting *value* for total height of masonry structure 砌体结构总高度限值

linear expansion coefficient 线膨胀系数

lintel 过梁

load bearing wall 承重墙

load-carrying capacity per bolt 单个普通螺栓承载能力

load-carrying capacity per high-strength holt 单个高强螺桂承载能力

load-carrying capacity per rivet 单个铆钉承载能力

log 原木

log timber structure 原木结构

long term rigidity of member 构件长期刚度

longitude horizontal bracing 纵向水平支撑

longitudinal steel bar 纵向钢筋

longitudinal stiffener 纵向加劲肋

longitudinal weld 纵向焊缝

losses of prestress 预应力损失

lump material 块体

M

main axis 强轴

main beam 主梁

major axis 强轴

manual welding 手工焊接

manufacture control 生产控制

map cracking 龟裂

masonry 砌体

masonry lintel 砖过梁

masonry member 无筋砌体构件

masonry units 块体

masonry-concrete structure 砖混结构

masonry-timber structure 砖木结构

mechanical properties of materials 材料力学性能

melt-thru 烧穿

method of sampling 抽样方法

minimum strength class of masonry 砌体材料最低强度等级

minor axle 弱轴
mix ratio of mortar 砂浆配合比
mixing water 拌合水
modified coefficient for allowable ratio of height to sectional thickness of masonry wall 砌体墙容许高厚比修正系数
modified coefficient of flexural strength for timber curved member 弧形木构件抗弯强度修正系数
modulus of elasticity of concrete 混凝土弹性模量
modulus of elasticity parallel to grain 顺纹弹性模量
moisture content 含水率
moment modified factor 弯矩调幅系数
monitor frame 天窗架
mortar 砂浆
multi-defense system of earthquake-resistant building 多道设防抗震建筑
multi-tube supported suspended structure 多筒悬挂结构

N

nailed joint 钉连接
net height 净高
net span 净跨度
net water/cement ratio 净水灰比
non-destructive inspection of weld 焊缝无损检验
non-destructive test 非破损检验
non-load-bearing wall 非承重墙
non-uniform cross-section beam 变截面梁
non-uniformly distributed strain coefficient of longitudinal tensile reinforcement 纵向受拉钢筋应变不均匀系数
normal concrete 普通混凝土
normal section 正截面
notch and tooth joint 齿连接
number of sampling 抽样数量

O

oblique section 斜截面
oblique-angle fillet weld 斜角角焊缝
one-way reinforced(or prestressed)concrete slab 单向板
open web roof truss 空腹屋架
ordinary concrete 普通混凝土
ordinary steel bar 普通钢筋
orthogonal fillet weld 直角角焊缝
outstanding width of flange 翼缘板外伸宽度

outstanding width of stiffener 加劲肋外伸宽度
over-all stability reduction coefficient of steel beam 钢梁整体稳定系数
overlap 焊瘤
overturning or slip resistance analysis 抗倾覆、滑移验算

P

padding plate 垫板
Partial penetrated butt weld 不焊透对接焊缝
Partition 非承重墙
penetrated butt weld 透焊对接焊缝
percentage of reinforcement 配筋率
perforated brick 多孔砖
pilastered wall 带壁柱墙
pit 凹坑
pith 髓心
plain concrete structure 素混凝土结构
plane hypothesis 平截面假定
plane structure 平面结构
plane trussed lattice grids 平面桁架系网架
plank 板材
plastic adoption coefficient of cross-section 截面塑性发展系数
plastic design of steel structure 钢结构塑性设计
plastic hinge 塑性铰
plasticity coefficient of reinforced concrete member in tensile zone 受拉区混凝土塑性影响系数
plate-like space frame 干板型网架
plate-like space truss 平板型网架
plug weld 塞焊缝
plywood 胶合板
plywood structure 胶合板结构
pockmark 麻面
polygonal top-chord roof truss 多边形屋架
post-tensioned prestressed concrete structure 后张法预应力混凝土结构
pre-cast reinforced concrete member 预制混凝土构件
prefabricated concrete structure 装配式混凝土结构
presetting time 初凝时间
prestressed concrete structure 预应力混凝土结构
prestressed steel structure 预应力钢结构
prestressed tendon 预应力筋
pre-tensioned prestressed concrete structure 先张法预应力混凝土结构

primary control 初步控制
production control 生产控制
properties of fresh concrete 可塑混凝土性能
properties of hardened concrete 硬化混凝土性能
property of building structural materials 建筑结构材料性能
purlin 檩条

Q

quality grade of structural timber 木材质量等级
quality grade of weld 焊缝质量级别
quality inspection of bolted connection 螺栓连接质量检验
quality inspection of masonry 砌体质量检验
quality inspection of riveted connection 铆钉连接质量检验
quasi-permanent *value* of live load on floor or roof 楼面、屋面活荷载准永久值

R

radial check 辐裂
ratio of axial compressive force to axial compressive ultimate capacity of section 轴压比
ratio of height to sectional thickness of wall or column 砌体墙柱高、厚比
ratio of reinforcement 配筋率
ratio of shear span to effective depth of section 剪跨比
redistribution of internal force 内力重分布
reducing coefficient of compressive strength in sloping grain for bolted connection 螺栓连接斜纹承压强度降低系数
reducing coefficient of liveload 活荷载折减系数
reducing coefficient of shearing strength for notch and tooth connection 齿连接抗剪强度降低系数
regular earthquake-resistant building 规则抗震建筑
reinforced concrete deep beam 混凝土深梁
reinforced concrete slender beam 混凝土浅梁
reinforced concrete structure 钢筋混凝土结构
reinforced masonry structure 配筋砌体结构
reinforcement ratio 配筋率
reinforcement ratio per unit volume 体积配筋率
relaxation of prestressed tendon 预应筋松弛
representative *value* of gravity load 重力荷载代表值
resistance to abrasion 耐磨性
resistance to freezing and thawing 抗冻融性
resistance to water penetration 抗渗性
reveal of reinforcement 露筋

right-angle filled weld 直角角焊缝
rigid analysis scheme 刚性方案
rigid connection 刚接
rigid transverse wall 刚性横墙
rigid zone 刚域
rigid-elastic analysis scheme 刚弹性方案
rigidity of section 截面刚度
rigidly supported continuous girder 刚性支座连续梁
ring beam 圈梁
rivet 铆钉
riveted connection 铆钉连接
riveted steel beam 铆接钢梁
riveted steel girder 铆接钢梁
riveted steel structure 铆接钢结构
roller support 滚轴支座
rolled steel beam 轧制型钢梁
roof board 屋面板
roof bracing system 屋架支撑系统
roof girder 屋面梁
roof plate 屋面板
roof slab 屋面板
roof system 屋盖
roof truss 屋架
rot 腐朽
round wire 光圆钢丝

S

safety classes of building structures 建筑结构安全等级
safety-bolt 保险螺栓
sapwood 边材
sawn lumber 方木
sawn timber structure 方木结构
saw-tooth joint failure 齿缝破坏
scarf joint 斜搭接
seamless steel pipe 无缝钢管
seamless steel tube 无缝钢管
second moment of area of transformed section 换算截面惯性矩
second order effect due to displacement 挠曲二阶效应
secondary axis 弱轴
secondary beam 次梁

section modulus of transformed section 换算截面模量
section steel 型钢
semi-automatic welding 半自动焊接
separated steel column 分离式钢柱
setting time 凝结时间
shake 环裂
shaped steel 型钢
shaped factor of wind load 风荷载体型系数
shear plane 剪面
shearing rigidity of section 截面剪变刚度
shearing stiffness of member 构件抗剪刚度
short stiffener 短加劲肋
short term rigidity of member 构件短期刚度
shrinkage 干缩
shrinkage of concrete 混凝干收缩
silos 贮仓
skylight truss 天窗架
slab 楼板
slab-column structure 板柱结构
slag inclusion 夹渣
sloping grain 斜纹
slump 坍落度
snow reference pressure 基本雪压
solid-web steel column 实腹式钢柱
space structure 空间结构
space suspended cable 悬索
spacing of bars 钢筋间距
spacing of rigid transverse wall 刚性横墙间距
spacing of stirrup legs 箍筋肢距
spacing of stirrups 箍筋间距
specified concrete 特种混凝土
spiral stirrup 螺旋箍筋
spiral weld 螺旋形焊缝
split ring joint 裂环连接
square pyramid space grids 四角锥体网架
stability calculation 稳定计算
stability reduction coefficient of axially loaded compression 轴心受压构件稳定系数
stair 楼梯
static analysis scheme of building 房屋静力计算方案

static design 房屋静力计算方案
statically determinate structure 静定结构
statically indeterminate structure 超静定结构
steel bar 钢筋
steel column component 钢柱分肢
steel column base 钢柱脚
steel fiber reinforced concrete structure 钢纤维混凝土结构
steel hanger 吊筋
steel mesh reinforced brick masonry member 方格网配筋砖砌体构件
steel pipe 钢管
steel plate 钢板
steel plate element 钢板件
steel strip 钢带
steel support 钢支座
steel tie 拉结钢筋
steel tie bar for masonry 砌体拉结钢筋
steel tube 钢管
steel tubular structure 钢管结构
steel wire 钢丝
stepped column 阶形柱
stiffener 加劲肋
stiffness of structural member 构件刚度
stiffness of transverse wall 横墙刚度
stirrup 箍筋
stone 石材
stone masonry 石砌体
stone masonry structure 石砌体结构
storey height 层高
straight-line joint failure 通缝破坏
straightness of structural member 构件平直线度
strand 钢绞线
strength classes of masonry units 块体强度等级
strength classes of mortar 砂浆强度等级
strength classes of structural steel 钢材强度等级
strength classes of structural timber 木材强度等级
strength classes(grades) of concrete 混凝土强度等级
strength classes(grades) of prestressed tendon 预应力筋强度等级
strength classes(grades) of steel bar 普通钢筋强度等级
strength of structural timber parallel to grain 木材顺纹强度

strong axis 强轴
structural system composed of bar 杆系结构
structural system composed of plate 板系结构
structural wall 结构墙
superposed reinforced concrete flexural member 叠合式混凝土受弯构件
suspended crossed cable net 双向正交索网结构
suspended structure 悬挂结构
swirl grain 涡纹

T

tensile(compressive) rigidity of section 截面拉伸(压缩)刚度
tensile(compressive) stiffness of member 构件抗拉(抗压)刚度
tensile(ultimate) strength of steel 钢材(钢筋)抗拉(极限)强度
test for properties of concrete structural members 构件性能检验
thickness of concrete cover 混凝土保护层厚度
thickness of mortar at bed joint 水平灰缝厚度
thin shell 薄壳
three hinged arch 三铰拱
tie bar 拉结钢筋
tie beam 系梁
tied framework 绑扎骨架
timber 木材
timber roof truss 木屋架
torsional-shear type of high-strength bolt 扭剪型高强度螺栓
torsional rigidity of section 截面扭转刚度
torsional stiffness of member 构件抗扭刚度
total breadth of structure 结构总宽度
total height of structure 结构总高度
total length of structure 结构总长度
transmission length of prestress 预应力传递长度
transverse horizontal bracing 横向水平支撑
transverse stiffener 横向加劲肋
transverse weld 横向焊缝
transversely distributed steel bar 横向分布钢筋
trapezoid roof truss 梯形屋架
triangular pyramid space grids 三角锥体网架
triangular roof truss 三角形屋架
trussed arch 橡架
trussed rafter 桁架拱
tube in tube structure 筒中筒结构

tube structure 筒体结构

twist 扭弯

two hinged arch 双铰拱

two sides(edges) supported plate 两边支承板

two-way reinforced(or prestressed) concrete slab 混凝土双向板

U

ultimate compressive strain of concrete 混凝土极限压应变

unbound prestressed concrete structure 无黏结预应力混凝土结构

undercut 咬边

uniform cross-section beam 等截面梁

unseasoned timber 湿材

upper flexible and lower rigid complex multistorey building 上柔下刚多层房屋

upper rigid lower flexible complex multistorey building 上刚下柔多层房屋

V

value of decompression prestress 预应力筋消压预应力值

value of effective prestress 预应筋有效预应力值

verification of serviceability limit states 正常使用极限状态验证

verification of ultimate limit states 承载能极限状态验证

vertical bracing 竖向支撑

visual examination of structural member 构件外观检查

visual examination of structural steel member 钢构件外观检查

visual examination of weld 焊缝外观检查

W

wall beam 墙梁

wall frame 壁式框架

wall-slab structure 墙板结构

warping 翘曲

warping rigidity of section 截面翘曲刚度

water retentivity of mortar 砂浆保水性

water tower 水塔

water/cement ratio 水灰比

weak axis 弱轴

weak region of earthquake-resistant building 抗震建筑薄弱部位

web plate 腹板

weld 焊缝

weld crack 焊接裂纹

weld defects 焊接缺陷

weld roof 焊根

weld toe 焊趾

weld ability of steel bar 钢筋可焊性
welded framework 焊接骨架
welded steel beam 焊接钢梁
welded steel girder 焊接钢梁
welded steel pipe 焊接钢管
welded steel structure 焊接钢结构
welding connection 焊缝连接
welding flux 焊剂
welding rod 焊条
welding wire 焊丝
wind fluttering factor 风振系数
wind reference pressure 基本风压
wind-resistant column 抗风柱
wood roof decking 屋面木基层
Y
yield strength(yield point) of steel 钢材(钢筋)屈服强度(屈服点)

REFERENCES

1. 教育部高教司. 大学英语课程教学要求. 中国大学教学, 2004(1).
2. 司显柱. 论新形势下的大学英语课程设置. 南昌航空工业学院学报, 2006(4).
3. 段兵廷主编. 土木工程专业英语. 武汉:武汉工业大学出版社, 2000.
4. 教育部土建英语教材编写组. 土建英语. 北京:高等教育出版社, 2000.
5. 建筑工程专业英语教程编写组. 建筑工程专业英语教程. 武汉:武汉工业大学出版社, 2000.
6. 俞戊孙. 建筑施工实用英语会话. 北京:中国建筑工业出版社, 1999.
7. 陆伟成. 建筑工程专业英语. 武汉:武汉理工大学出版社, 2007.
8. 孙爱荣. 建筑工程专业英语. 哈尔滨:哈尔滨工业大学出版社, 2005.
9. 吴承霞. 建筑工程专业英语. 北京:北京大学出版社, 2009.
10. Des Moines, Lowa. *Step-by-Step Basic Masonry and Concrete*. New York: Meredith Corporation, 2003:70.
11. 建筑施工日常英语 800 句, http://wenku.baidu.com/view/7c602ba20029bd64783e2cb7.html.
12 工程管理实用英语对话, http://wenku.baidu.com/view/8ffbe645be1e650e52ea99fa.html.